LYNCHBURG COLLEGE

SYMPOSIUM READINGS

CLASSICAL SELECTIONS ON GREAT ISSUES

SERIES TWO
VOLUME I

MAN AND THE UNIVERSE

Bacon
Butterfield
Copernicus
Galileo
Newton
Faraday
Einstein
Heisenberg
Jeans

UNIVERSITY
PRESS OF
AMERICA

SERIES TWO
SYMPOSIUM READINGS
Lynchburg College in Virginia

Compiled and Edited by the
following faculty members of Lynchburg College:

Kenneth E. Alrutz, Ph.D., University of Pennsylvania; Assistant
Professor of English

Virginia B. Berger, M.A., Harvard University; Associate Professor
of Music

Anne Marshall Bippus, Ed.D., University of Virginia; Associate
Professor of Education

James L. Campbell, Ph.D., University of Virginia; Associate Professor of English

Robert L. Frey, Ph.D., University of Minnesota; Professor of
History

James A. Huston, Ph.D., New York University; Dean of the College,
Professor of History and International Relations

Shannon McIntyre Jordan, Ph.D., University of Georgia; Instructor
in Philosophy

Jan G. Linn, D.Min., Christian Theological Seminary; Assistant
Professor, College Chaplain

Peggy S. Pittas, M.A., Dalhousie University; Associate Professor
of Psychology

Clifton W. Potter, Jr., Ph.D., University of Virginia; Professor
of History

Julius A. Sigler, Ph.D., University of Virginia; Professor of
Physics

Phillip H. Stump, Ph.D., University of California at Los Angeles;
Assistant Professor of History

Thomas C. Tiller, Ph.D., Florida State University; Dean of Student
Affairs, Professor of Education

Copyright © 1982 by

University Press of America, Inc.

P.O. Box 19101, Washington, D.C. 20036

Library of Congress Catalog Card Number: 81-71466

ACKNOWLEDGEMENTS

The following copyrighted materials have been used with the permission of the copyright holders:

From *The New Organon* by Francis Bacon. c. 1960, Bobbs-Merrill Educational Publishing.

From *Three Copernican Treatises*, "Commentariolus" by Nicholas Copernicus. c.1959; From *Dialogues Concerning Two New Sciences* by Galileo. c.1954; Dover Publications, Inc.

From *Discoveries and Opinions of Galileo*, "The Starry Messenger", "Letter to the Grand Duchess" by Galileo. c.1957, Doubleday & Co., Inc.

From *Out of My Later Years*, "Science and Religion" by Albert Einstein. c 1950, Estate of Albert Einstein.

From *Physics and Philosophy* by Werner Heisenberg. c 1958, Harper & Row.

From *Principia Mathematicas* by Isaac Newton, c.1952, University of California Press.

The authors acknowledge with appreciation the permissions granted by these holders of the respective copyrights.

INTRODUCTION TO THE SERIES

These reading selections are offered in support of the Senior Symposium at Lynchburg College.

The Symposium is intended to be a kind of capstone for the general education programs of the College. It has the general purpose of serving as a "bridge" between formal college courses and the continuing study and consideration of major issues by responsible college graduates. The objectives are to develop a sense of perspective in recognizing that mankind has faced similar issues through the ages and that no panaceas exist for settling human problems once for all; to bring together the thinking of people of broad experience, varying points of view, and diverse backgrounds in seeking ways to meet the questions which demand attention in public affairs, and to develop an awareness of unity in knowledge and of the interrelationships of problems in various fields of activity and interest.

In a way, the course may be said to have had four major historic antecedents - "paternal and maternal grandparents" as it were. These were (1) the senior course in "Moral Philosophy" common in American colleges in the nineteenth century; (2) "Chapel" or "College Assembly" common in many American colleges until fairly recently; (3) the "Great Books" programs of Columbia, Chicago, and Harvard, and (4) the "Great Issues" course at Dartmouth, University of Denver, Purdue, and Michigan State.

In *A History of Curriculum* written for the Carnegie Council on Policy Studies in Higher Education, Frederick Rudolph describes the first of these as follows:

> The senior course in moral philosophy justified the curriculum, it rationalized it. It asserted the unity of knowledge, sent the young graduates out into the world with a reassuring sense of their own fitness to play a role in upholding the moral order; it treated them like men, they who had thus far been treated like boys. By the 19th Century the moral philosophy course, Aristotelian in origin and English and Scottish in modification, had become a remarkable excursion into social and individual ethics. Politics, eco-

vii

nomics, sociology, law, government, history,
esthetics, international law, and fine arts
were territories into which the moral philosophers...roamed.
. . . . .

Indeed, the senior course in moral philosophy
was the last moment when the idea of a liberal education as an expression of the unity
of knowledge could be honestly held.

The traditional "chapel" evolved from an essentially religious exercise to a series of lectures on a
wide range of topics in the arts and sciences for which
the attendance of all students commonly was required.
At Lynchburg College, this later became "assembly". At
first, attendance was required of all; then grades and
credit were assigned according to attendance. It survived in the catalogue as an assembly for which the
College might require attendance at certain meetings
until 1972, though actually it had fallen into disuse
for want of a satisfactory meeting site.

The "Great Books" program at Columbia inspired the
one at the University of Chicago where it became a major element of the undergraduate curriculum, and then
reached the ultimate at St. John's, Annapolis, where
"Great Books" became the basis for the entire curriculum. Similar programs developed at Harvard, Notre
Dame, and other institutions across the country. All
these doubtless owed a great deal to the publication of
The Harvard Classics in 1909 and 1910. As a Harvard committee, reporting on the organization of its course on
"Selected Great Books", stated:

The aim of such a course would be the full-
est understanding of the work read rather
than of men or periods represented, crafts-
manship evinced, historic or literary development shown, or anything else. These
other matters would be admitted only inso-
far as they are necessary to allow the work
to speak for itself.

The treatment which is attempted by these
great themes can only do its best to be
worthy of them. They themselves are its
inspirations... The instructor can only
seek to be a means by which the authors
teach the course.

viii

The distinguished jurist, Learned Hand, once wrote:

> I venture to believe that it is as important
> to a judge called upon to pass on a question
> of constitutional law, to have at least a
> bowing acquaintance with Thucydides,...Shake-
> speare, with Machiavelli, Plato, and Bacon,
> as with the books which have been specifical-
> ly written on the subject.
>
> For in such matters everything turns upon the
> spirit in which he approaches the questions
> before him... Men do not gather figs of this-
> tles, nor supply institutions from judges
> whose outlook is limited by parish or class.

The Dartmouth course on "Great Issues" was a week-
ly assembly of seniors to hear visiting speakers on
current issues. But for background reading, the students depended on *The New York Times* and other periodicals.

The Senior Symposium is an attempt to bring all
these elements together. The approach is to undertake
the discussion of major, continuing issues on the basis
of current ideas presented by visiting scholars, public
officials, artists, business and professional people,
and other leaders of thought and opinion with the perspective of ideas presented in "great books".

Theodore M. Greene of Yale University has written
that the four basic ingredients of a liberal education
- what the college should give to every student - are
the following:

1. Training in the accurate and felicitous
   use of *language* as the essential condition of all reflection, self-expression,
   and communication with others.

2. Training in the acquisition of *factual
   knowledge* of ourselves, our society, and
   other societies, the physical world,
   and ultimate reality so far as it is hu-
   manly knowable.

3. Training in mature and responsible *eval-
   uation and decision* in the controversial
   areas of social policy, morality, art,
   and religion.

4. Training in *synoptic comprehension,* i.e.,
   in the escape from multiple provincial-
   isms which bedevil mankind, and in the
   attainment of larger and more inclusive
   perspectives.

As conceived here, the purposes of the Senior Symposium fall precisely into the third and fourth of
these categories.

The "great issues" for consideration in the course
are taken to be broad areas of great, continuing problems of mankind. Within each area certain questions
are taken up for consideration in relation to the
broader aspects of the problems.

Thus for the first semester the general themes are
chosen from "The Nature of Man," "Education: Means and
Ends," "Freedom and Tyranny," "Poverty and Wealth," and
"War and Peace." Current questions taken up include
such matters as individual freedom and internal secur-
ity, the "new look" in United States foreign and mili-
tary policies, economic stablization, and foreign aid
and trade. In the second semester the general themes
are "Man and the Universe," "Science, Technology and
Society," "Man and the Imagination: The Fine Arts,"
"Faith and Morals," and "Man and Society."

The procedure includes assigned readings in selections from the "classics," lectures each week, usually
by visiting authorities, on some related current question or to present contemporary views on broader issues,
and then group discussions later in the week on the
readings and lectures. Each student, then, has a week-
ly reading assignment, one lecture, and one discussion
session.

In the discussion sections, the students express
and test their own views on the basis of what they have
read and heard, not only in the course but in their
total experience. No effort is made to arrive at general conclusions or "final answers" for any of the
questions. Each student is encouraged to develop his
own ideas and to keep his mind open to modify his views
whenever additional evidence demands it.

There is no assumption that the Symposium represents any thorough coverage of all the fields that it
touches. As the capstone to a student's general education, it gains its depth by drawing upon all the educa-

tional experience of the student and by dwelling on particular questions. In discussions cutting across the lines of formal disciplines, students have a chance to learn a great deal from each other around a focus of concern common to all. It also is a demonstration for seniors that their education is far from complete, that much remains to be learned, and learning is a process which must continue through their lives.

While these readings are intended primarily for seniors, many of them, of course, would be appropriate for any level. Indeed, a common definition of a classic is a work which has such a quality of timelessness that it speaks to all ages of persons as well as to all ages of history. Still we have found that the discussion of these works is much more fruitful for seniors than for underclassmen. It is to be expected and hoped that students will have encountered many of these works earlier in their academic careers. But here they will be approaching them with a different perspective, and with the benefit of added experience and maturity which will make their reading and discussion all the more meaningful and all the more delightful.

The temptation in making such a collection as this is to include too much. A weakness of many anthologies which grows out of this is to select such small segments that the reader misses many of the significant points and especially the well reasoned argument of the masters. In this series some short selections have been included for a specific point, but for the most part selections comprise whole chapters or several chapters of the works. The plan has been deliberately to include substantially more selections than may be used in a semester. This allows flexibility in making assignments from year to year, and it provides additional material upon which students may draw for the preparation of special reports or simply to extend their reading.

Readings are groups in units which are intended to correspond to assignments. For example, in one semester all five themes might be used with three units for each. In another, only three themes might be used with five assignments for each. Or there may be other combinations suitable for the purposes of the instructors. In further reading one might explore a particular theme in greater depth, or might search for relationships among readings which happen to be grouped under different themes.

This selection of readings is the result of the efforts of the group of Lynchburg College faculty members whose names appear on the title page. With the support of a National Endowment for the Humanities pilot grant, they worked individually, in small teams, and as a group to choose the readings and to prepare study questions and brief introductory statements. The questions, intended mainly to guide the reading, may also be useful for examinations and quizzes, but especially for discussion. Introductions have been left short that "the books may speak for themselves."

Dr. Jordan and Dr. Potter did further editing of materials; Dr. Potter did the final editing; and Shirley W. Moore prepared the materials for publication. George R. White, Jr., Assistant Librarian for Instructional Services, served as library consultant.

J.A.H.

INTRODUCTION TO VOLUME I

From the beginning of the process we call thought, man has longed to know his place in the cosmos. This need to understand the environment has been the driving force behind the growth of science, which is but a particular way of knowing. Although the appearance of nature as manifested in the shape of a continent, the elusiveness of a breeze, or the violence of a black hole is temporal, the physical laws governing each are unchanging and apply throughout the universe. Science has provided a sense of order and, with understanding, power.

For the past three hundred years, science has grown in breadth, depth, and respect. The readings in this theme were chosen to provide a sense of the way in which scientific thought differs from other forms of intellectual activity, to provide a sense of how our vision of the universe has changed with the development of modern science, and to acquaint one with the mainstreams of twentieth-century thought.

CONTENTS

Francis Bacon, THE NEW ORGANON

1. What did Bacon find to be inadequate in earlier philosophy? Did he feel that the inadequacies could be corrected?

2. Define the scientific method as Bacon sees it. What role does experimentation play in it? What does Bacon mean by true induction?

3. What wider effects does Bacon expect from the use of his method? What do you feel are the legitimate expectations which we can have concerning science in our own time?

4. What are the limitations of our senses and our minds in affording us true knowledge of the universe? How can we overcome these limitations according to Bacon? What are the other most important obstacles to scientific investigation?

Born in England, in 1561, Sir Francis Bacon (1561-1626) was a man well-acquainted with power and with scandal. Luckily his political career and eventual ruin are all but forgotten thanks to his lasting reputation as a scholar. His *New Organon*, written in 1620, six years before his death, was one of the great manifestos of the scientific revolution. Aristotle's *Organon* was a set of logical writings which set forth the Aristotelian methodology; Bacon's *New Organon* was at once a rejection of Aristotle and an effort to replace his methodology with a new scientific method.

Those who have taken upon them to lay down the law of nature as a thing already searched out and understood, whether they have spoken in simple assurance or professional affectation, have therein done philosophy and the sciences great injury. For as they have been successful in inducing belief, so they have been effective in quenching and stopping inquiry; and have done more harm by spoiling and putting an end to other men's efforts than good by their own. Those on the other hand who have taken a contrary course, and asserted that absolutely nothing can be known—whether it were from hatred of the ancient sophists, or from uncertainty and fluctuation of mind, or even from a kind of fullness of learning, that they fell upon this opinion—have certainly advanced reasons for it that are not to be despised; but yet they have neither started from true principles nor rested in the just conclusion, zeal and affectation having carried them much too far. The more ancient of the Greeks (whose writings are lost) took up with better judgment a position between these two extremes—between the presumption of pronouncing on everything, and the despair of comprehending anything; and though frequently and bitterly complaining of the difficulty of inquiry and the obscurity of things, and like impatient horses champing at the bit, they did not the less follow up their object and engage with nature, thinking (it seems) that this very question—viz., whether or not anything can be known—was to be settled not by arguing, but by trying. And yet they too, trusting entirely to the force of their understanding, applied no rule, but made everything turn upon hard thinking and perpetual working and exercise of the mind.

Now my method, though hard to practice, is easy to explain; and it is this. I propose to establish progressive stages of certainty. The evidence of the sense, helped and guarded by a certain process of correction, I retain. But the mental operation which follows the act of sense I for the most part reject; and instead of it I open and lay out a new and certain path for the mind to proceed in, starting directly from the simple sensuous perception. The necessity of this was felt, no doubt, by those who attributed so much importance to logic, showing thereby that they were in search of helps for the understanding, and had no confidence in the native and spontaneous process of the mind. But this remedy comes too late to do any good, when the mind is already, through the daily intercourse and conversation of life, occupied with unsound doctrines and beset on all sides by vain imaginations. And therefore that art of logic, coming (as I said) too late to the rescue, and no way able to set matters right again, has had the effect of fixing errors rather than disclosing truth. There remains but one course for the recovery of a sound and healthy condition—namely, that the entire work of the understanding be commenced afresh, and the mind itself be from the very outset not left to take its own course, but guided at every step; and the business be done as if by machinery. Certainly if in things mechanical men had set to work with their naked hands, without help or force of instruments, just as in things intellectual they have set to work with little else than the naked forces of the understanding, very small would the matters have been which, even with their best efforts applied in conjunction, they could have attempted or accomplished. Now (to pause a while upon this example and look in it as in a glass) let us suppose that some vast obelisk were (for the decoration of a triumph or some such magnificence) to be removed from its place, and that men should set to work upon it with their naked hands, would not any sober spectator think them mad? And if they should then send for more people, thinking that in that way they might manage it, would he not think them all the madder? And if they then proceeded to make a selection, putting away the weaker hands, and using only the strong and vigorous, would he not think them madder than ever? And if lastly, not content with this, they resolved to call in aid the art of athletics, and required all their men to come with hands, arms, and sinews well anointed and medicated according to the rules of the art, would he not cry out that they were only taking pains to show a kind of method and discretion in their madness? Yet just so it is that men proceed in matters intellectual—with just the same kind of mad effort and useless combination of forces—when they hope great things either from the number and cooperation or from the excellency and acuteness of individual wits; yea, and when they endeavor by logic (which may be considered as a kind of athletic art) to strengthen the sinews of the understanding, and yet with all this study and endeavor it is apparent to any true judgment that they are but applying the naked intellect all the time; whereas in every great work to be done by the hand of man it is manifestly impossible, without instruments and machinery, either for the strength of each to be exerted or the strength of all to be united.

Upon these premises two things occur to me of which, that they may not be overlooked, I would have men reminded. First, it falls out fortunately as I think for the allaying of contradictions and heartburnings, that the honor and reverence due to the ancients remains untouched and undiminished, while I may carry out my designs and at the same time reap the fruit of my modesty. For if I should profess that I, going the same road as the ancients, have something better to produce, there must needs have been some comparison or rivalry between us (not to be avoided by any art of words) in respect of excellency or ability of wit; and though in this there would be nothing unlawful or new (for if there be anything misapprehended by them, or falsely laid down, why may not I, using a liberty common to all, take exception to it?) yet the contest, however just and allowable, would have been an unequal one perhaps, in respect of the measure of my own powers. As it is, however (my object being to open a new way for the understanding, a way by them untried and unknown), the case is altered: party zeal and emulation are at an end, and I appear merely as a guide to point out the road—an of-

4

fice of small authority, and depending more upon a kind of luck than upon any ability or excellency. And thus much relates to the persons only. The other point of which I would have men reminded relates to the matter itself.

Be it remembered then that I am far from wishing to interfere with the philosophy which now flourishes, or with any other philosophy more correct and complete than this which has been or may hereafter be propounded. For I do not object to the use of this received philosophy, or others like it, for supplying matter for disputations or ornaments for discourse—for the professor's lecture and for the business of life. Nay, more, I declare openly that for these uses the philosophy which I bring forward will not be much available. It does not lie in the way. It cannot be caught up in passage. It does not flatter the understanding by conformity with preconceived notions. Nor will it come down to the apprehension of the vulgar except by its utility and effects.

Let there be therefore (and may it be for the benefit of both) two streams and two dispensations of knowledge, and in like manner two tribes or kindreds of students in philosophy—tribes not hostile or alien to each other, but bound together by mutual services; let there in short be one method for the cultivation, another for the invention, of knowledge.

And for those who prefer the former, either from hurry or from considerations of business or for want of mental power to take in and embrace the other (which must needs be most men's case), I wish that they may succeed to their desire in what they are about, and obtain what they are pursuing. But if there be any man who, not content to rest in and use the knowledge which has already been discovered, aspires to penetrate further; to overcome, not an adversary in argument, but nature in action; to seek, not pretty and probable conjectures, but certain and demonstrable knowledge—I invite all such to join themselves, as true sons of knowledge, with me, that passing by the outer courts of nature, which numbers have trodden, we may find a way at length into her inner chambers. And to make my meaning clearer and to familiarize the thing by giving it a name, I have chosen to call one of these methods or ways *Anticipation of the Mind,* the other *Interpretation of Nature.*

Moreover, I have one request to make. I have on my own part made it my care and study that the things which I shall propound should not only be true, but should also be presented to men's minds, how strangely soever preoccupied and obstructed, in a manner not harsh or unpleasant. It is but reasonable, however (especially in so great a restoration of learning and knowledge), that I should claim of men one favor in return, which is this: if anyone would form an opinion or judgment either out of his own observation, or out of the crowd of authorities, or out of the forms of demonstration (which have now acquired a sanction like that of judicial laws), concerning these speculations of mine, let him not hope that he can do it in passage or by the by; but let him examine the thing thoroughly; let him make some little trial for himself of the way which I describe and lay out; let him familiarize his thoughts with that subtlety of nature to which experience bears witness; let him correct by seasonable patience and due delay the depraved and deep-rooted habits of his mind; and when all this is done and he has begun to be his own master, let him (if he will) use his own judgment.

APHORISMS

[BOOK ONE]

I

Man, being the servant and interpreter of Nature, can do and understand so much and so much only as he has observed in fact or in thought of the course of nature. Beyond this he neither knows anything nor can do anything.

II

Neither the naked hand nor the understanding left to itself can effect much. It is by instruments and helps that the work is done, which are as much wanted for the understanding as for the hand. And as the instruments of the hand either give motion or guide it, so the instruments of the mind supply either suggestions for the understanding or cautions.

III

Human knowledge and human power meet in one; for where the cause is not known the effect cannot be produced. Nature to be commanded must be obeyed; and that which in contemplation is as the cause is in operation as the rule.

IV

Toward the effecting of works, all that man can do is to put together or put asunder natural bodies. The rest is done by nature working within.

7

V

The study of nature with a view to works is engaged in by the mechanic, the mathematician, the physician, the alchemist, and the magician; but by all (as things now are) with slight endeavor and scanty success.

VI

It would be an unsound fancy and self-contradictory to expect that things which have never yet been done can be done except by means which have never yet been tried.

VII

The productions of the mind and hand seem very numerous in books and manufactures. But all this variety lies in an exquisite subtlety and derivations from a few things already known, not in the number of axioms.

VIII

Moreover, the works already known are due to chance and experiment rather than to sciences; for the sciences we now possess are merely systems for the nice ordering and setting forth of things already invented, not methods of invention or directions for new works.

IX

The cause and root of nearly all evils in the sciences is this —that while we falsely admire and extol the powers of the human mind we neglect to seek for its true helps.

X

The subtlety of nature is greater many times over than the subtlety of the senses and understanding; so that all those specious meditations, speculations, and glosses in which men indulge are quite from the purpose, only there is no one by to observe it.

XI

As the sciences which we now have do not help us in finding out new works, so neither does the logic which we now have help us in finding out new sciences.

XII

The logic now in use serves rather to fix and give stability to the errors which have their foundation in commonly received notions than to help the search after truth. So it does more harm than good.

XIII

The syllogism is not applied to the first principles of sciences, and is applied in vain to intermediate axioms, being no match for the subtlety of nature. It commands assent therefore to the proposition, but does not take hold of the thing.

XIV

The syllogism consists of propositions, propositions consist of words, words are symbols of notions. Therefore if the notions themselves (which is the root of the matter) are confused and overhastily abstracted from the facts, there can be no firmness in the superstructure. Our only hope therefore lies in a true induction.

XV

There is no soundness in our notions, whether logical or physical. Substance, Quality, Action, Passion, Essence itself, are not sound notions; much less are Heavy, Light, Dense, Rare, Moist, Dry, Generation, Corruption, Attraction, Repulsion, Element, Matter, Form, and the like; but all are fantastical and ill defined.

XVI

Our notions of less general species, as Man, Dog, Dove, and of the immediate perceptions of the sense, as Hot, Cold, Black, White, do not materially mislead us; yet even these are sometimes confused by the flux and alteration of matter and the mixing of one thing with another. All the others which men have hitherto adopted are but wanderings, not being abstracted and formed from things by proper methods.

XVII

Nor is there less of willfulness and wandering in the construction of axioms than in the formation of notions, not excepting even those very principles which are obtained by common induction; but much more in the axioms and lower propositions educed by the syllogism.

XVIII

The discoveries which have hitherto been made in the sciences are such as lie close to vulgar notions, scarcely beneath the surface. In order to penetrate into the inner and further recesses of nature, it is necessary that both notions and axioms be derived from things by a more sure and guarded way, and that a method of intellectual operation be introduced altogether better and more certain.

XIX

There are and can be only two ways of searching into and discovering truth. The one flies from the senses and particulars to the most general axioms, and from these principles, the truth of which it takes for settled and immovable, proceeds to judgment and to the discovery of middle axioms. And this way is now in fashion. The other derives axioms from the senses and particulars, rising by a gradual and unbroken ascent, so that it arrives at the most general axioms last of all. This is the true way, but as yet untried.

XX

The understanding left to itself takes the same course (namely, the former) which it takes in accordance with logical order. For the mind longs to spring up to positions of higher generality, that it may find rest there, and so after a little while wearies of experiment. But this evil is increased by logic, because of the order and solemnity of its disputations.

XXI

The understanding left to itself, in a sober, patient, and grave mind, especially if it be not hindered by received doctrines, tries a little that other way, which is the right one, but with little progress, since the understanding, unless directed and assisted, is a thing unequal, and quite unfit to contend with the obscurity of things.

XXII

Both ways set out from the senses and particulars, and rest in the highest generalities; but the difference between them is infinite. For the one just glances at experiment and particulars in passing, the other dwells duly and orderly among them.

11

The one, again, begins at once by establishing certain abstract and useless generalities, the other rises by gradual steps to that which is prior and better known in the order of nature.

XXIII

There is a great difference between the Idols of the human mind and the Ideas of the divine. That is to say, between certain empty dogmas, and the true signatures and marks set upon the works of creation as they are found in nature.

XXIV

It cannot be that axioms established by argumentation should avail for the discovery of new works, since the subtlety of nature is greater many times over than the subtlety of argument. But axioms duly and orderly formed from particulars easily discover the way to new particulars, and thus render sciences active.

XXV

The axioms now in use, having been suggested by a scanty and manipular experience and a few particulars of most general occurrence, are made for the most part just large enough to fit and take these in; and therefore it is no wonder if they do not lead to new particulars. And if some opposite instance, not observed or not known before, chance to come in the way, the axiom is rescued and preserved by some frivolous distinction; whereas the truer course would be to correct the axiom itself.

XXVI

The conclusions of human reason as ordinarily applied in matters of nature, I call for the sake of distinction *Anticipations of Nature* (as a thing rash or premature). That reason which is elicited from facts by a just and methodical process, I call *Interpretation of Nature.*

XXVII

Anticipations are a ground sufficiently firm for consent, for even if men went mad all after the same fashion, they might agree one with another well enough.

XXVIII

For the winning of assent, indeed, anticipations are far more powerful than interpretations, because being collected from a few instances, and those for the most part of familiar occurrence, they straightway touch the understanding and fill the imagination; whereas interpretations, on the other hand, being gathered here and there from very various and widely dispersed facts, cannot suddenly strike the understanding; and therefore they must needs, in respect of the opinions of the time, seem harsh and out of tune, much as the mysteries of faith do.

XXIX

In sciences founded on opinions and dogmas, the use of anticipations and logic is good; for in them the object is to command assent to the proposition, not to master the thing.

XXX

Though all the wits of all the ages should meet together and combine and transmit their labors, yet will no great progress ever be made in science by means of anticipations; because radical errors in the first concoction of the mind are not to be cured by the excellence of functions and subsequent remedies.

XXXI

It is idle to expect any great advancement in science from the superinducing and engrafting of new things upon old. We must begin anew from the very foundations, unless we would revolve forever in a circle with mean and contemptible progress.

XXXII

The honor of the ancient authors, and indeed of all, remains untouched, since the comparison I challenge is not of wits or faculties, but of ways and methods, and the part I take upon myself is not that of a judge, but of a guide.

XXXIII

This must be plainly avowed: no judgment can be rightly formed either of my method or of the discoveries to which it leads, by means of anticipations (that is to say, of the reasoning which is now in use); since I cannot be called on to abide by the sentence of a tribunal which is itself on trial.

XXXIV

Even to deliver and explain what I bring forward is no easy matter, for things in themselves new will yet be apprehended with reference to what is old.

XXXV

It was said by Borgia of the expedition of the French into Italy, that they came with chalk in their hands to mark out their lodgings, not with arms to force their way in. I in like manner would have my doctrine enter quietly into the minds that are fit and capable of receiving it; for confutations cannot be employed when the difference is upon first principles and very notions, and even upon forms of demonstration.

XXXVI

One method of delivery alone remains to us which is simply this: we must lead men to the particulars themselves, and their series and order; while men on their side must force themselves for a while to lay their notions by and begin to familiarize themselves with facts.

XXXVII

The doctrine of those who have denied that certainty could be attained at all has some agreement with my way of proceeding at the first setting out; but they end in being infinitely separated and opposed. For the holders of that doctrine assert simply that nothing can be known. I also assert that not much can be known in nature by the way which is now in use. But then they go on to destroy the authority of the senses and understanding; whereas I proceed to devise and supply helps for the same.

XXXVIII

The idols and false notions which are now in possession of the human understanding, and have taken deep root therein, not only so beset men's minds that truth can hardly find entrance, but even after entrance is obtained, they will again in the very instauration of the sciences meet and trouble us, unless men being forewarned of the danger fortify themselves as far as may be against their assaults.

XXXIX

There are four classes of Idols which beset men's minds. To these for distinction's sake I have assigned names, calling

But of these several kinds of Idols I must speak more largely and exactly, that the understanding may be duly cautioned.

XLV

The human understanding is of its own nature prone to suppose the existence of more order and regularity in the world than it finds. And though there be many things in nature which are singular and unmatched, yet it devises for them parallels and conjugates and relatives which do not exist. Hence the fiction that all celestial bodies move in perfect circles, spirals and dragons being (except in name) utterly rejected. Hence too the element of fire with its orb is brought in, to make up the square with the other three which the sense perceives. Hence also the ratio of density of the so-called elements is arbitrarily fixed at ten to one. And so on of other dreams. And these fancies affect not dogmas only, but simple notions also.

XLVI

The human understanding when it has once adopted an opinion (either as being the received opinion or as being agreeable to itself) draws all things else to support and agree with it. And though there be a greater number and weight of instances to be found on the other side, yet these it either neglects and despises, or else by some distinction sets aside and rejects, in order that by this great and pernicious predetermination the authority of its former conclusions may remain inviolate. And therefore it was a good answer that was made by one who, when they showed him hanging in a temple a picture of those who had paid their vows as having escaped shipwreck, and would have him say whether he did not now acknowledge the power of the gods—"Aye," asked he again, "but where are they painted that were drowned after their vows?" And such is the way of all superstition, whether in astrology, dreams, omens, divine judgments, or the like;

wherein men, having a delight in such vanities, mark the events where they are fulfilled, but where they fail, though this happen much oftener, neglect and pass them by. But with far more subtlety does this mischief insinuate itself into philosophy and the sciences; in which the first conclusion colors and brings into conformity with itself all that come after, though far sounder and better. Besides, independently of that delight and vanity which I have described, it is the peculiar and perpetual error of the human intellect to be more moved and excited by affirmatives than by negatives; whereas it ought properly to hold itself indifferently disposed toward both alike. Indeed, in the establishment of any true axiom, the negative instance is the more forcible of the two.

XLVII

The human understanding is moved by those things most which strike and enter the mind simultaneously and suddenly, and so fill the imagination; and then it feigns and supposes all other things to be somehow, though it cannot see how, similar to those few things by which it is surrounded. But for that going to and fro to remote and heterogeneous instances by which axioms are tried as in the fire, the intellect is altogether slow and unfit, unless it be forced thereto by severe laws and overruling authority.

XLVIII

The human understanding is unquiet; it cannot stop or rest, and still presses onward, but in vain. Therefore it is that we cannot conceive of any end or limit to the world, but always as of necessity it occurs to us that there is something beyond. Neither, again, can it be conceived how eternity has flowed down to the present day, for that distinction which is commonly received of infinity in time past and in time to come can by no means hold; for it would thence follow that one infinity is greater than another, and that infinity is wasting

19

away and tending to become finite. The like subtlety arises touching the infinite divisibility of lines, from the same inability of thought to stop. But this inability interferes more mischievously in the discovery of causes; for although the most general principles in nature ought to be held merely positive, as they are discovered, and cannot with truth be referred to a cause, nevertheless the human understanding being unable to rest still seeks something prior in the order of nature. And then it is that in struggling toward that which is further off it falls back upon that which is nearer at hand, namely, on final causes, which have relation clearly to the nature of man rather than to the nature of the universe; and from this source have strangely defiled philosophy. But he is no less an unskilled and shallow philosopher who seeks causes of that which is most general, than he who in things subordinate and subaltern omits to do so.

XLIX

The human understanding is no dry light, but receives an infusion from the will and affections; whence proceed sciences which may be called "sciences as one would." For what a man had rather were true he more readily believes. Therefore he rejects difficult things from impatience of research; sober things, because they narrow hope; the deeper things of nature, from superstition; the light of experience, from arrogance and pride, lest his mind should seem to be occupied with things mean and transitory; things not commonly believed, out of deference to the opinion of the vulgar. Numberless, in short, are the ways, and sometimes imperceptible, in which the affections color and infect the understanding.

L

But by far the greatest hindrance and aberration of the human understanding proceeds from the dullness, incompetency, and deceptions of the senses; in that things which strike the

20

sense outweigh things which do not immediately strike it, though they be more important. Hence it is that speculation commonly ceases where sight ceases; insomuch that of things invisible there is little or no observation. Hence all the working of the spirits enclosed in tangible bodies lies hid and unobserved of men. So also all the more subtle changes of form in the parts of coarser substances (which they commonly call alteration, though it is in truth local motion through exceedingly small spaces) is in like manner unobserved. And yet unless these two things just mentioned be searched out and brought to light, nothing great can be achieved in nature, as far as the production of works is concerned. So again the essential nature of our common air, and of all bodies less dense than air (which are very many), is almost unknown. For the sense by itself is a thing infirm and erring; neither can instruments for enlarging or sharpening the senses do much; but all the truer kind of interpretation of nature is effected by instances and experiments fit and apposite; wherein the sense decides touching the experiment only, and the experiment touching the point in nature and the thing itself.

LI

The human understanding is of its own nature prone to abstractions and gives a substance and reality to things which are fleeting. But to resolve nature into abstractions is less to our purpose than to dissect her into parts; as did the school of Democritus, which went further into nature than the rest. Matter rather than forms should be the object of our attention, its configurations and changes of configuration, and simple action, and law of action or motion; for forms are figments of the human mind, unless you will call those laws of action forms.

LII

Such then are the idols which I call *Idols of the Tribe,* and which take their rise either from the homogeneity of the sub-

21

stance of the human spirit, or from its preoccupation, or from its narrowness, or from its restless motion, or from an infusion of the affections, or from the incompetency of the senses, or from the mode of impression.

LIII

The *Idols of the Cave* take their rise in the peculiar constitution, mental or bodily, of each individual; and also in education, habit, and accident. Of this kind there is a great number and variety. But I will instance those the pointing out of which contains the most important caution, and which have most effect in disturbing the clearness of the understanding.

LIV

Men become attached to certain particular sciences and speculations, either because they fancy themselves the authors and inventors thereof, or because they have bestowed the greatest pains upon them and become most habituated to them. But men of this kind, if they betake themselves to philosophy and contemplation of a general character, distort and color them in obedience to their former fancies; a thing especially to be noticed in Aristotle, who made his natural philosophy a mere bond servant to his logic, thereby rendering it contentious and well-nigh useless. The race of chemists, again out of a few experiments of the furnace, have built up a fantastic philosophy, framed with reference to a few things; and Gilbert also, after he had employed himself most laboriously in the study and observation of the loadstone, proceeded at once to construct an entire system in accordance with his favorite subject.

LV

There is one principal and as it were radical distinction between different minds, in respect of philosophy and the sciences, which is this: that some minds are stronger and apter to mark the differences of things, others to mark their resemblances. The steady and acute mind can fix its contemplations and dwell and fasten on the subtlest distinctions; the lofty and discursive mind recognizes and puts together the finest and most general resemblances. Both kinds, however, easily err in excess, by catching the one at gradations, the other at shadows.

LVI

There are found some minds given to an extreme admiration of antiquity, others to an extreme love and appetite for novelty; but few so duly tempered that they can hold the mean, neither carping at what has been well laid down by the ancients, nor despising what is well introduced by the moderns. This, however, turns to the great injury of the sciences and philosophy, since these affectations of antiquity and novelty are the humors of partisans rather than judgments; and truth is to be sought for not in the felicity of any age, which is an unstable thing, but in the light of nature and experience, which is eternal. These factions therefore must be abjured, and care must be taken that the intellect be not hurried by them into assent.

LVII

Contemplations of nature and of bodies in their simple form break up and distract the understanding, while contemplations of nature and bodies in their composition and configuration overpower and dissolve the understanding, a distinction well seen in the school of Leucippus and Democritus as compared with the other philosophies. For that school is so busied with the particles that it hardly attends to the structure, while the others are so lost in admiration of the structure that they do not penetrate to the simplicity of nature. These kinds of contemplation should therefore be alternated and taken by turns, so that the understanding may be ren-

23

dered at once penetrating and comprehensive, and the inconveniences above mentioned, with the idols which proceed from them, may be avoided.

LVIII

Let such then be our provision and contemplative prudence for keeping off and dislodging the *Idols of the Cave,* which grow for the most part either out of the predominance of a favorite subject, or out of an excessive tendency to compare or to distinguish, or out of partiality for particular ages, or out of the largeness or minuteness of the objects contemplated. And generally let every student of nature take this as a rule: that whatever his mind seizes and dwells upon with peculiar satisfaction is to be held in suspicion, and that so much the more care is to be taken in dealing with such questions to keep the understanding even and clear.

LIX

But the *Idols of the Market Place* are the most troublesome of all—idols which have crept into the understanding through the alliances of words and names. For men believe that their reason governs words; but it is also true that words react on the understanding; and this it is that has rendered philosophy and the sciences sophistical and inactive. Now words, being commonly framed and applied according to the capacity of the vulgar, follow those lines of division which are most obvious to the vulgar understanding. And whenever an understanding of greater acuteness or a more diligent observation would alter those lines to suit the true divisions of nature, words stand in the way and resist the change. Whence it comes to pass that the high and formal discussions of learned men end oftentimes in disputes about words and names; with which (according to the use and wisdom of the mathematicians) it would be more prudent to begin, and so by means of definitions reduce them to order. Yet even definitions cannot cure this evil in dealing with natural and material things, since the definitions themselves consist of words, and those words beget others. So that it is necessary to recur to individual instances, and those in due series and order, as I shall say presently when I come to the method and scheme for the formation of notions and axioms.

LX

The idols imposed by words on the understanding are of two kinds. They are either names of things which do not exist (for as there are things left unnamed through lack of observation, so likewise are there names which result from fantastic suppositions and to which nothing in reality corresponds), or they are names of things which exist, but yet confused and ill-defined, and hastily and irregularly derived from realities. Of the former kind are Fortune, the Prime Mover, Planetary Orbits, Element of Fire, and like fictions which owe their origin to false and idle theories. And this class of idols is more easily expelled, because to get rid of them it is only necessary that all theories should be steadily rejected and dismissed as obsolete.

But the other class, which springs out of a faulty and unskillful abstraction, is intricate and deeply rooted. Let us take for example such a word as *humid* and see how far the several things which the word is used to signify agree with each other, and we shall find the word *humid* to be nothing else than a mark loosely and confusedly applied to denote a variety of actions which will not bear to be reduced to any constant meaning. For it both signifies that which easily spreads itself round any other body; and that which in itself is indeterminate and cannot solidize; and that which readily yields in every direction; and that which easily divides and scatters itself; and that which easily unites and collects itself; and that which readily flows and is put in motion; and that which readily clings to another body and wets it; and that which is easily reduced to a liquid, or being solid easily melts. Accordingly, when you

25

come to apply the word, if you take it in one sense, flame is humid; if in another, air is not humid; if in another, fine dust is humid; if in another, glass is humid. So that it is easy to see that the notion is taken by abstraction only from water and common and ordinary liquids, without any due verification.

There are, however, in words certain degrees of distortion and error. One of the least faulty kinds is that of names of substances, especially of lowest species and well-deduced (for the notion of *chalk* and of *mud* is good, of *earth* bad); a more faulty kind is that of actions, as *to generate, to corrupt, to alter;* the most faulty is of qualities (except such as are the immediate objects of the sense) as *heavy, light, rare, dense,* and the like. Yet in all these cases some notions are of necessity a little better than others, in proportion to the greater variety of subjects that fall within the range of the human sense.

LXI

But the *Idols of the Theater* are not innate, nor do they steal into the understanding secretly, but are plainly impressed and received into the mind from the playbooks of philosophical systems and the perverted rules of demonstration. To attempt refutations in this case would be merely inconsistent with what I have already said, for since we agree neither upon principles ncr upon demonstrations there is no place for argument. And this is so far well, inasmuch as it leaves the honor of the ancients untouched. For they are no wise disparaged—the question between them and me being only as to the way. For as the saying is, the lame man who keeps the right road outstrips the runner who takes a wrong one. Nay, it is obvious that when a man runs the wrong way, the more active and swift he is, the further he will go astray.

But the course I propose for the discovery of sciences is such as leaves but little to the acuteness and strength of wits, but places all wits and understandings nearly on a level. For as in the drawing of a straight line or a perfect circle, much depends on the steadiness and practice of the hand, if it be done by aim of hand only, but if with the aid of rule or compass, little or nothing; so is it exactly with my plan. But though particular confutations would be of no avail, yet touching the sects and general divisions of such systems I must say something; something also touching the external signs which show that they are unsound; and finally something touching the causes of such great infelicity and of such lasting and general agreement in error; that so the access to truth may be made less difficult, and the human understanding may the more willingly submit to its purgation and dismiss its idols.

2

Herbert Butterfield, THE ORIGINS OF MODERN SCIENCE

1. Why did the scientific revolution occur in the fields of physics and astronomy rather than in other fields?

2. According to Aristotle's theory of motion, why does a heavy object fall to earth faster than a lighter one? Why does this theory demand that the earth be the center of the universe?

3. Why was it so difficult to overcome the views of Aristotle and Ptolemy? What was the point of greatest weakness in Aristotle's theory of motion?

Herbert Butterfield (1900-   ), British historian and Regius Professor at Cambridge has written books on a wide range of topics. This work, *The Origins of Modern Science*, has become almost a classic introduction to the history of the rise of modern science.

THE HISTORICAL IMPORTANCE OF A
THEORY OF IMPETUS

It is one of the paradoxes of the whole story with which we have to deal that the most sensational step leading to the scientific revolution in astronomy was taken long before the discovery of the telescope—even long before the Danish astronomer, Tycho Brahé, in the latter part of the sixteenth century, had shown the great improvement that it was still possible to achieve in observations made with the naked eye. When William Harvey in England opened up new paths for physiology by his study of the action of the heart, he alluded once or twice to his use of a magnifying glass, but he carried out his revolutionary work before any serviceable kind of microscope had become available. With regard to the transformation of the science of mechanics, it is remarkable to what an extent even Galileo discusses the ordinary phenomena of everyday life, conjectures what would happen if a stone were thrown from the mast of a moving ship, or plays with pellets on inclined planes in a manner that had long been customary. In fact, we shall find that in both celestial and terrestrial physics—which hold the strategic place in the whole movement—change is brought about, not by new observations or additional evidence in the first instance, but by transpositions that were taking place inside the minds of the scientists themselves. In this connection it is not irrelevant to note that of all forms of mental activity the most difficult to induce, even in the minds of the young who may be presumed not to have lost their flexibility, is the art of handling the same bundle of data as before, but placing them in a new system of relations with one another by giving them a different framework, all of which virtually means putting on a different kind of thinking-cap for the moment. It is easy to teach anybody a new fact about Richelieu, but it needs light from heaven to enable a teacher to break the old framework in which the student has been accustomed to seeing his Richelieu—the framework which is built up sometimes far too rigidly by the Higher Certificate student, and into which he will fit whatever new information he ever afterwards acquires on this subject. But the supreme paradox of the scientific revolution is the fact that things which we find it easy to instil into boys at school, because we see that they start off on the right foot—things which would strike us as the ordinary natural way of looking at the universe, the obvious way of regarding the behaviour of falling bodies, for example—defeated the greatest intellects for centuries, defeated Leonardo da Vinci and at the marginal point even Galileo, when their minds were wrestling on the very frontiers of human thought with these very problems. Even the great geniuses who broke through the ancient views in some special field of study—Gilbert, Bacon and Harvey, for example—would remain stranded in a species of medievalism when they went outside that chosen field. It required their combined efforts to clear up certain simple things which we should now regard as obvious to any unprejudiced mind, and even easy for a child.

A particular development of ideas which was already taking place in the later middle ages has come to stand as the first chapter in the history of the transition to what we call the scientific revolution. It is a field of thought upon which an expositor can embark only with the greatest trepidation, in view of the vicissitudes of lecturers at the very beginning of modern times. Students of history will remember how the humanists of the Renaissance, Erasmus included, were accustomed to complaining of the boredom—deriding the sophistries and subtleties—of the scholastic lectures which they had to endure at the university. Occasionally they specified the forms of teaching and lecturing to which they most objected, and as they particularly mentioned those discussions of mechanics with which we have now to concern ourselves, it will no doubt be prudent to make the examination of their teaching as brief as possible. It is curious that these despised scholastic disquisitions should now have come to hold a remarkable key-position in the story of the evolution of the modern mind. Perhaps the lack of mathematics, or the failure to think of mathematical ways of formulating things, was partly responsible for what appeared to be verbal subtleties and an excessive straining of language in these men who were almost yearning to find the way to the modern science of mechanics.

Of all the intellectual hurdles which the human mind has been faced with and has overcome in the last fifteen hundred years, the one which seems to me to have been the most amazing in character and the most stupendous in the scope of its consequences is the one relating to the problem of motion—the one which was not quite disposed of by Galileo, though it received a definitive form of settlement shortly after his time in the full revised statement of what every schoolboy learns to call the law of inertia. On this question of motion the Aristotelian teaching, precisely because it carried such an intricate dovetailing of observations and explanations —that is to say, precisely because it was part of a system which was such a colossal intellectual feat in itself—was hard for the human mind to escape from, and gained a strong hold on medieval scholastic thought. Furthermore, it remains as the essential background of the story—it continues to present the presiding issue—until the time of Galileo himself; in other words, until the first half of the seventeenth century. On the Aristotelian theory all heavy terrestrial bodies had a natural motion towards the centre of the universe, which for medieval thinkers was at or near the centre of the earth; but motion in any other direction was violent motion, because it contradicted the ordinary tendency of a body to move to what was regarded as its natural place. Such motion depended on the operation of a mover, and the Aristotelian doctrine of inertia was a doctrine of rest—it was motion, not rest, that always required to be explained. Wherever this motion existed, and however long it existed, something had to be brought in to account for it.

The essential feature of this view was the assertion or the presumption that a body would keep in movement only so long as a mover was actually in contact with it, imparting motion to it all the time. Once the mover ceased to operate, the movement stopped— the body fell straight to earth or dropped suddenly to rest. Further —a point that will seem very heretical to the present day—it was argued that, provided the resistance of the medium through which the body passed remained a constant, the speed of the body would be proportionate to what we should describe as the force consistently being exerted upon it by the mover. A constant force exerted by the mover over a given length of time produced not any acceleration at all, but a uniform motion for the whole period. On the other hand, if there was any variation in the resistance of the medium— the difference between moving in air and moving in water, for example—the speed would vary in inverse proportion to this, provided the other factors remained constant. And if the resistance were reduced to nought, the speed would be infinite; that is to say, if the movement took place in a vacuum, bodies would move from one place to another instantaneously. The absurdity of this was one

31

of the reasons why the Aristotelians regarded a complete void as impossible, and said that God Himself could not make one.

It is astonishing to what a degree not only this theory but its rivals—even the ones which superseded it in the course of the scientific revolution—were based on the ordinary observation of the data available to common sense. And, as writers have clearly pointed out, it is not relevant for us to argue that if the Aristotelians had merely watched the more carefully they would have changed their theory of inertia for the modern one—changed over to the view that bodies tend to continue either at rest or in motion along a straight line until something intervenes to stop them or deflect their course. It was supremely difficult to escape from the Aristotelian doctrine by merely observing things more closely, especially if you had already started off on the left foot and were hampered beforehand with the whole system of interlocking Aristotelian ideas. In fact, the modern law of inertia is not the thing you would discover by mere photographic methods of observation—it required a different kind of thinking-cap, a transposition in the mind of the scientist himself; for we do not actually see ordinary objects continuing their rectilinear motion in that kind of empty space which Aristotle said could not exist, and sailing away to that infinity which also he said could not possibly exist; and we do not in real life have perfectly spherical balls moving on perfectly smooth horizontal planes—the trick lay in the fact that it occurred to Galileo to imagine these. Furthermore, even when men were coming extraordinarily near to what we should call the truth about local motion, they did not clinch the matter— the thing did not come out clear and clean—until they had realised and had made completely conscious to themselves the fact that they were in reality transposing the question into a different realm— they were discussing not real bodies as we actually observe them in the real world, but geometrical bodies moving in a world without resistance and without gravity—moving in that boundless emptiness of Euclidean space which Aristotle had regarded as unthinkable. In the long run, therefore, we have to recognise that here was a problem of a fundamental nature, and it could not be solved by close observation within the framework of the older system of ideas—it required a transposition in the mind.

As often happened with such theories in those days, if not now, the Aristotelian doctrine of motion might seem to correspond in a self-evident manner with most of the data available to common sense, but there were small pockets of fact which did not square with the theory at the first stage of the argument; they were unamenable to the Aristotelian laws at what we should call the ordinary common-sense level. There were one or two anomalies which required a further degree of analysis before they could be satisfactorily adjusted to the system; and perhaps, as some writers have said, the Aristotelian theory came to a brilliant peak in the manner by which it hauled these exceptional cases into the synthesis and established (at a second remove) their conformity with the stated rules. On the argument so far as we have taken it, an arrow ought to have fallen to the ground the moment it lost contact with 'the bow-string; for neither the bow-string nor anything else could impart a motion which would continue after the direct contact with the original mover had been broken. The Aristotelians explained the continued movement of projectiles by the commotion which the initial movement had produced in the air—especially as the air which was being pushed and compressed in front had to rush round behind to prevent that vacuum which must never be allowed to take place. At this point in the argument there even occurred a serious fault in observation which harassed the writers on physical science for many centuries. It was thought that the rush of air produced an actual initial acceleration in the arrow after it had left the bow-string, and it is curious to note that Leonardo da Vinci and later writers shared this mistake—the artillerymen of the Renaissance were victims of the same error—though there had been people in the later middle ages who had taken care not to commit themselves on this point. The motion of a projectile, since it was caused by a disturbance in the medium itself, was a thing which it was not possible to imagine taking place in a vacuum.

Furthermore, since the Aristotelian commentators held something corresponding to the view that a constant uniform force only produced uniform motion, there was a second serious anomaly to be explained—it was necessary to produce special reasons to account for the fact that falling bodies were observed to move at an accelerating speed. Once again the supporters of the older teaching used the argument from the rush of air, or the increasing pressure, as there was a greater height of atmosphere above, while as the body descended there would be a diminishing column of air below, and therefore a diminishing resistance. Alternatively they used Aristotle's argument that the falling body moved more jubilantly every moment because it found itself nearer home.

From the fourteenth to the seventeenth century, then, this Aristotelian doctrine of motion persisted in the face of recurrent controversy, and it was only in the later stages of that period that the satisfactory alternative emerged, somewhat on the policy of picking up the opposite end of the stick. Once this question was solved in the modern manner, it altered much of one's ordinary thinking about the world and opened the way for a flood of further discoveries and reinterpretations, even in the realm of common sense, before any very elaborate experiments had been embarked upon. It was as though science or human thought had been held up by a barrier until this moment—the waters dammed because of an initial defect in one's attitude to everything in the universe that had any sort of motion—and now the floods were released; change and discovery were bound to come in cascades even if there were no other factors working for a scientific revolution. Indeed, we might say that a change in one's attitude to the movement of things that move was bound to result in so many new analyses of various kinds of motion that it constituted a scientific revolution in itself.

Apart from all this there was one special feature of the problem which made the issue momentous. We have not always brought home to ourselves the peculiar character of that Aristotelian universe in which the things that were in motion had to be accompanied by a mover all the time. A universe constructed on the mechanics of Aristotle had the door half-way open for spirits already; it was a universe in which unseen hands had to be in constant operation, and sublime Intelligences had to roll the planetary spheres around. Alternatively, bodies had to be endowed with souls and aspirations, with a "disposition" to certain kinds of motion, so that matter itself seemed to possess mystical qualities. The modern law of inertia, the modern theory of motion, is the great factor which in the seventeenth century helped to drive the spirits out of the world and opened the way to a universe that ran like a piece of clockwork. Not only so—but the very first men who in the middle ages launched the great attack on the Aristotelian theory were conscious of the fact that this colossal issue was involved in the question. The first of the important figures, Jean Buridan, in the middle of the fourteenth century pointed out that his alternative interpretation would eliminate the need for the Intelligences that turned the celestial spheres. He even noted that the Bible provided no authority for these spiritual agencies—they were demanded by the teaching of the ancient Greeks, not by the Christian religion as such. Not much later than this, Nicholas of Oresme went farther still, and said that on the new alternative theory God might have started off the universe as a kind of clock and left it to run of itself.

Ever since the earlier years of the twentieth century at latest, therefore, a great and growing interest has been taken in that school of thinkers who so far back as the fourteenth century were challenging the Aristotelian explanations of motion, and who put forward an alternative doctrine of "impetus" which—though imperfect in itself—must represent the first stage in the history of the scientific revolution. And if it is imagined that this kind of argument falls into one of the traps which it is always necessary to guard against —picking out from the middle ages mere anticipations and casual analogies to modern ideas—the answer to that objection will be clear to us if we bear in mind the kind of rules that ought to govern historians in such matters. Here we have a case of a consistent body of teaching carried on and developed as a tradition by a school of thinkers particularly in Paris, and still being taught in Paris at the beginning of the sixteenth century. It has a continuous history— we know how this teaching passed into Italy, how it was in general resisted even in the days of the humanists, how Leonardo da Vinci picked it up, however, and how some of what were once considered to be remarkable strokes of modernity, remarkable flashes of genius, in his notebooks, were in reality transcriptions from fourteenth-century Parisian scholastic writers. We know how the teaching gained a foothold in later sixteenth-century Italy, amongst the men who influenced Galileo, how it was misunderstood on occasion— sometimes only partially appropriated—and how the early writings of Galileo on motion belonged precisely to this school of teaching, being based on that doctrine of the "impetus" which it is our purpose to examine. It is even known fairly certainly in what edition Galileo read the words of certain writers belonging to this fourteenth-century Parisian school. Indeed, Galileo could have produced much, though not quite all, that we find in his juvenile works on this particular subject if he had lived in the fourteenth century; and in this field one might very well ask what the world with its Renaissance and so forth had been doing in the meantime. It has been suggested that if printing had been invented two centuries earlier the doctrine of "impetus" would have produced a more rapid general development in the history of science, and would not have needed so long to pass from the stage of Jean Buridan to the stage of Galileo.

If the orthodox doctrine of the middle ages had been based on Aristotle, however, it has to be noted that, both then and during the Renaissance (as well as later still), the attacks on Aristotle—the theory of impetus included—would themselves be based on some ancient thinker. Here we touch on one of the generative factors, not only in the formation of the modern world, but also in the development of the scientific revolution—namely, the discovery of the fact that even Aristotle had not reigned unchallenged in the ancient days—all of which resulted in a healthy friction, resulting in the emergence of important problems which the middle ages had to make up their own minds about, so that men were driven to some kind of examination of the workings of nature themselves, even if only because they had to make up their minds between Aristotle and some rival teacher. It also appears that a religious factor affected the rise of that movement which produced the theory of impetus, and in a curious manner, which one tries in vain to analyse away, a religious taboo operated for once in favour of freedom for scientific hypothesis. In the year 1277 a council in Paris condemned a large number of Aristotelian theses, such as the view that even God could not create a void, or an infinite universe, or a plurality of worlds; and that decision—one in which certain forms of partisanship were involved—was apparently extended by the Archbishop of Canterbury to this country. The religions that came within the orbit of these decisions must have been the seat of a certain anti-Aristotelian bias already; and certainly from this time both Paris and Oxford showed remarkable developments in this direction in the field of what we should call physical science. From this time the discussion of the possibility of the existence of empty space, or of an infinite universe, or of a plurality of worlds makes a remarkable step forward in Paris. And amongst the names concerned in this development are some which figure in the rise of the doctrine of impetus. It has also been pointed out that in the same Parisian tradition there was a tendency towards something in the nature of mathematical physics, though the mathematics of the time were not sufficiently advanced to allow of this being carried very far or to produce anything like the achievement of Galileo in the way of a mathematical approach to scientific problems. We must avoid the temptation, however, to stress unduly the apparent analogies with modern times and the "anticipations" which are so easy to discover in the past—things which often owe a little, no doubt, to the trick-mirrors of the historian. And though it may be useful sometimes, in order to illustrate a point, we must beware of submitting to the fascination of "what might have been."

The people who chiefly concern us, then, are the fourteenth-century writers, Jean Buridan, Albert of Saxony and Nicholas of Oresme, but we are interested in them chiefly as leaders of a tradition which still existed in Paris at the beginning of the sixteenth century. They are important for other things besides their teaching on the subject of impetus, because though the contemporaries of Erasmus laughed at their Parisian lecturers for discussing not only "uniform motion" and "difform motion", but also "uniform difform motion"—all carried to a great degree of subtlety—it transpired in the sixteenth century, when the world was looking for a formula to represent the uniform acceleration of falling bodies, that the solution of the problem had been at their disposal for a long time in the medieval formula for the case of uniformly difform motion. The whole development which we are studying took place amongst people who, in fact, were working upon questions and answers which had been suggested by Aristotle. These people came up against the Aristotelian theory of motion at the very points where we should expect the attack to take place, namely, in connection with the two particularly doubtful questions of the movement of projectiles and the acceleration of falling bodies. If we take a glance at the kind of arguments they used, we are not only watching the kind of critical procedure which would take place even in the middle ages in connection with the current forms of orthodox Aristotelianism, we are really observing the early stages of the great debate on the issues that lay at the heart of the scientific revolution itself. Indeed, the arguments which were employed at this early period often reappeared—with reference to precisely the same instances—even in the major works of Galileo, for they passed into general currency in the course of time. And if they seem simple arguments based on the ordinary phenomena available to common sense, we ought to remember that many of the newer arguments brought forward by Galileo himself at a later stage of the story were really of the same type.

It was the view of this Parisian school that the projectile was carried forward by an actual impetus which it had acquired, and which bodies were capable of acquiring, from the mere fact of being in motion; and this impetus was supposed to be a thing inside the body itself—occasionally it was described as an impetuosity that had been imparted to it; occasionally you see it discussed as though it were itself movement which the body acquired as the result of being in motion; but in any case it made it possible for men to contemplate the continued motion of a body after the contact with the original mover had been lost. It was explained that the impetus lay in the body and continued there, as heat stays in a red-hot poker after it has been taken from the fire; while in the case of falling bodies the effect was described as accidental gravity, an additional gravity which the body acquired as a result of being in motion, so that the acceleration of falling bodies was due to the effects of impetus being continually added to the constant fall due to ordinary weight. A constant force exerted on a body, therefore, produced not uniform motion but a uniform rate of acceleration. It is to be noted, however, that Leonardo da Vinci, like a number of others who accepted the general theory of impetus, failed to follow the Parisian school in the application of their teaching to the acceleration of falling bodies. Whereas the Aristotelians thought that falling bodies rushed more quickly as they got nearer home, the new teaching inverted this, and said that it was rather the distance from the starting-point that mattered. If two bodies fell to earth along the same line BC, the one which had started higher up at A would move more quickly from B to C than the one that started at B, though in this particular part of their course they were both equally distant from the centre of the earth. It followed from the new doctrine that if a cylindrical hole were cut through the earth, passing through the centre, a body, when it reached the centre, would be carried forward on its own impetus for some distance, and indeed would oscillate about the centre for some time—a thing impossible to conceive under the terms of the ancient theory. There was a further point in regard to which the Aristotelians had been unconvincing; for if the continued flight of a projectile were really due not to the thrower but to the rush of air, it was difficult to see why the air should carry a stone so much farther than a ball of feathers—why one should be able to throw the stone the greater distance. The newer school showed that, starting at a given pace, a greater impetus would be communicated to the stone by reason of the density of its material than to a feather;

though, of course, a larger body of the same material would not travel farther—a large stone would not be more easy to throw than a small one. Mass was used as the measure of the impetus which corresponded with a given speed.

Since Aristotle found it necessary on occasion to regard the air as a resisting factor, he was open to the charge that one could not then—in the next breath, so to speak—start using the argument that the air was also the actual propellant. The new school said that the air could not be the propellant except in the case of a high wind; and they brought the further objection that if the original perturbation of the air—the rush which occurred when the bow-string started the arrow—had the capacity to repeat itself, pushing the arrow on and on, there could be no reason why it should ever stop; it ought to go on for ever repeating itself, communicating further perturbations to every next region of the atmosphere. Furthermore, a thread tied to a projectile ought to be blown ahead of it, instead of trailing behind. In any case, on the Aristotelian view of the matter it ought to be impossible for an arrow to fly against the wind. Even the apostles of the new theory of impetus, however, regarded a projectile as moving in a straight line until the impetus had exhausted itself, and then making a direct vertical drop to earth. They looked upon this impetus as a thing which gradually weakened and wore itself out, just as a poker grows cold when taken from the fire. Or, said Galileo, it was like the reverberations which go on in a bell long after it has been struck, but which gradually fade away. Only, in the case of the celestial bodies and the orbs which carried the planets round the sky, the impulse never exhausted itself—the pace of these bodies never slackened since there was no air-resistance to slow them down. Therefore, it could be argued, God might have given these things their initial impetus, and their motion would then continue for ever.

The theory of the impetus did not solve all problems, however, and proved to be only the half-way house to the modern view, which is fairly explicit in Galileo though it received its perfect formulation only in Descartes—the view that a body continues its motion in a straight line until something intervenes to halt or slacken or deflect it. As I have already mentioned, this modern law of inertia is calculated to present itself more easily to the mind when a transposition has taken place—when we see, not real bodies, moving under the restrictions of the real world and clogged by the atmosphere, but geometrical bodies sailing away in empty Euclidean space. Archimedes, whose works were more completely discovered at the Renaissance and became very influential especially after the translation published in 1543, appears to have done something to assist and encourage this habit of mind; and nothing could have been more important than the growing tendency to geometrise or mathematise a problem. Nothing is more effective, after people have long been debating and wrangling and churning the air, than the appearance of a person who draws a line on the blackboard, which with the help of a little geometry solves the whole problem in an instant. In any case, it is possible that Archimedes, who taught people to think of the weight of a thing in water and then its weight in air and finally, therefore, its weight when unencumbered by either, helped to induce some men to pick up the problem of motion from the opposite end to the usual one, and to think of the simplest form of motion as occurring when there was no resisting medium to complicate it. So you assumed the tendency in bodies to continue their existing motion along a straight line, and you set about afterwards to examine the things which might clog or hamper or qualify that motion; whereas Aristotle, assuming that the state of rest was natural and that bodies tended to return to it when left to themselves, had the difficult task of providing an active mover that should operate as long as the body continued to have any movement at all.

On the other hand, it may be true to say that Aristotle, when he thought of motion, had in mind a horse drawing a cart, so that his whole feeling for the problem was spoiled by his preoccupation with a misleading example. The very fact that his teaching on the subject of projectiles was so unsatisfactory may have helped to produce the phenomenon of a later age which, when it thought of motion, had rather the motion of projectiles in mind, and so acquired a different feeling in regard to the whole matter.

THE CONSERVATISM OF COPERNICUS

An introductory sketch of the medieval view of the cosmos must be qualified first of all with the reservation that in this particular realm of thought there were variations, uncertainties, controversies and developments which it would obviously be impossible to describe in detail. On the whole, therefore, it would be well, perhaps, if we were to take Dante's view of the universe as a pattern, because it will be easy to note in parenthesis some of the important variations that occurred, and at the same time this policy will enable us to see in a single survey the range of the multiple objections which it took the Copernican theory something like a hundred and fifty years to surmount.

According to Dante, what one must have in mind is a series of spheres, one inside another, and at the heart of the whole system lies the motionless earth. The realm of what we should call ordinary matter is confined to the earth and its neighbourhood—the region below the moon; and this matter, the stuff that we can hold between our fingers and which our modern physical sciences set out to study, is humble and unstable, being subject to change and decay for reasons which we shall examine later. The skies and the heavenly bodies—the rotating spheres and the stars and planets that are attached to them—are made of a very tangible kind of matter too, though it is more subtle in quality and it is not subject to change and corruption. It is not subject to the physical laws that govern the more earthy kind of material which we have below the moon. From the point of view of what we should call purely physical science, the earth and the skies therefore were cut off from one another and were separate organisations for a medieval student, though in a wider system of thought they dovetailed together to form one coherent cosmos.

As to the ordinary matter of which the earth is composed, it is formed of four elements, and these are graded according to their virtue, their nobility. There is earth, which is the meanest stuff of all, then water, then air and, finally, fire, and this last comes highest

41

in the hierarchy. We do not see these elements in their pure and undiluted form, however—the earthy stuff that we handle when we
pick up a little soil is a base compound and the fire that we actually
see is mixed with earthiness. Of the four elements, earth and water
possess gravity; they have a tendency to fall; they can only be at rest
at the centre of the universe. Fire and air do not have gravity, but
possess the very reverse; they are characterised by levity, an actual
tendency to rise, though the atmosphere clings down to the earth because it is loaded with base mundane impurities. For all the elements
have their spheres, and aspire to reach their proper sphere, where
they find stability and rest; and when even flame has soared to its
own upper region it will be happy and contented, for here it can be
still and can most endure. If the elements did not mix—if they were
all at home in their proper spheres—we should have a solid sphere
of earth at the heart of everything and every particle of it would be
still. We should then have an ocean covering that whole globe, like
a cap that fitted all round, then a sphere of air, which far above
mountain-tops was supposed to swirl round from east to west in
sympathy with the movement of the skies. Finally, there would come
the region of enduring fire, fitting like a sphere over all the rest.

That, however, would be a dead universe. In fact, it was a corol-
lary of this whole view of the world that ordinary motion up or
down or in a straight line could only take place if there was something wrong—something displaced from its proper sphere. It mattered very much, therefore, that the various elements were not all
in order but were mixed and out of place—for instance, some of
the land had been drawn out above the waters, raised out of its
proper sphere at the bottom, to provide habitable ground. On this
land natural objects existed and, since they were mixtures, they
might, for example, contain water, which as soon as it was released would tend to seek its way down to the sea. On the other
hand, they might contain the element of fire, which would come out
of them when they burned and would flutter and push its way upwards, aspiring to reach its true home. But the elements are not
always able to follow their nature in this pure fashion—occasionally
the fire may strike downwards, as in lightning, or the water may
rise in the form of vapour to prepare a store of rain. On one point,
however, the law was fixed: while the elements are out of their proper
spheres they are bound to be unstable—there cannot possibly be
restfulness and peace. Woven, as we find them, on the surface of the

42

globe, they make a mixed and chancy world, a world that is subject to constant mutation, liable to dissolution and decay.

It is only in the northern hemisphere that land emerges, pro-
truding above the waters that cover the rest of the globe. This land
has been pulled up, out of its proper sphere, says Dante—drawn
not by the moon or the planets or the ninth sky, but by an influence
from the fixed stars, in his opinion. The land stretches from the
Pillars of Hercules in the west to the Ganges in the east, from the
Equator in the south to the Arctic Circle in the north. And in the
centre of this whole habitable world is Jerusalem, the Holy City.
Dante had heard stories of travellers who had found a great deal
more of the continent of Africa, found actual land much farther south
than he had been taught to consider possible. As a true rationalist
he seems to have rejected "fables" that contradicted the natural
science of his time, remembering that travellers were apt to be liars.
The disproportionate amount of water in the world and the un-
balanced distribution of the land led to some discussion of the
whereabouts of the earth's real centre. The great discoveries, however, culminating in the unmistakable discovery of America, provoked certain changes in ideas, as well as a debate concerning the
possibility of the existence of inhabited countries at the antipodes.
There was a growing view that earth and water, instead of coming
in two separate circles, the one above the other, really dovetailed
into one another to form a single sphere.

All this concerns the sublunary region; but there is another realm
of matter to be considered, and this, as we have already seen, comes
under a different polity. The skies are not liable to change and decay, for they—with the sun, the stars and the planets—are formed
of a fifth element, an incorruptible kind of matter, which is subject
to a different set of what we should call physical laws. If earth tends
to fall to the centre of the universe, and fire tends to rise to its
proper sphere above the air itself, the incorruptible stuff that forms
the heavens has no reason for discontent—it is fixed in its congenial
place already. Only one motion is possible for it—namely, circular
motion—it must turn while remaining in the same place. According to Dante there are ten skies, only the last of them, the Empyrean
Heaven, the abode of God, being at rest. Each of the skies is a sphere
that surrounds the globe of the earth, and though all these spheres
are transparent they are sufficiently tangible and real to carry one
or more of the heavenly bodies round on their backs as they rotate

about the earth—the whole system forming a set of transparent spheres, one around the other, with the hard earth at the centre of all. The sphere nearest to the earth has the moon attached to it, the others carry the planets or the sun, until we reach the eighth, to which all the fixed stars are fastened. A ninth sphere has no planet or star attached to it, nothing to give visible signs of its existence; but it must be there, for it is the *primum mobile*—it turns not only itself but all the other spheres or skies as well, from east to west, so that once in twenty-four hours the whole celestial system wheels round the motionless earth. This ninth sphere moves more quickly than any of the others, for the spirits which move it have every reason to be ardent. They are next to the Empyrean Heaven.

In the system of Aristotle the spheres were supposed to be formed of a very subtle æthereal substance, moving more softly than liquids and without any friction; but with the passage of time the idea seems to have become coarsened and vulgarised. The successive heavens turned into glassy or crystalline globes, solid but still transparent, so that it became harder for men to keep in mind the fact that they were frictionless and free from weight, though the Aristotelian theory in regard to these points was still formally held.

The original beauty of this essentially Aristotelian system had been gravely compromised, however, by the improvements which had been made in astronomical observation since the time when it had been given its original shape; for even in the ancient world astronomy afforded a remarkable example of the progress which could be achieved in science by the sheer passage of time—the accumulating store of observations and the increasing precision in the recordings. Early in the Christian era, in the age of Ptolemy, the complications had become serious, and in the middle ages both the Arabs and the Christians produced additions to the intricacies of the system. The whole of the celestial machinery needed further elaboration to account for planets which, as viewed by the observer, now stopped in the sky, now turned back on their courses, now changed their distance from the earth, now altered their speed. However irregular the motion of the planets might seem to be, however curious the path that they traced, their behaviour must somehow or other be reduced to circular, even uniform circular motion—if necessary to a complicated series of circular motions each corrective of the other. Dante explains how Venus goes round with the sphere which forms the third of the skies, but as this does not quite correspond to the phenomena, another sphere which revolves independently is fixed to the sphere of the third sky, and the planet rides on the back of the smaller sphere (sitting like a jewel there, says Dante), reflecting the light of the sun. But writers varied on this point, and we meet the view that the planet was rather like a knot in a piece of wood, or represented a mere thickening of the material that formed the whole celestial sphere—a sort of swelling that caught the sunlight and shone with special brilliance as a result.

Writers differed also on the question whether the whole of the more elaborate machinery—the eccentrics or epicycles—as devised by Ptolemy and his successors, really existed in the actual architecture of the skies, though the theory of the crystalline spheres persisted until the seventeenth century. Since the whole complicated system demanded eighty spheres, some of which must apparently intersect one another as they turned round, some writers regarded the circles and epicycles as mere geometrical devices that formed a basis for calculation and prediction. And some men who believed that the nine skies were genuine crystalline spheres might regard the rest of the machinery as a mathematical way of representing those irregularities and anomalies which they knew they were unable properly to explain. In any case, it was realised long before the time of Copernicus that the Ptolemaic system, in spite of all its complications, did not exactly cover the phenomena as observed. In the sixteenth and seventeenth centuries we shall still find people who will admit that the Ptolemaic system is inadequate, and who will say that a new one must be discovered, though for understandable reasons they reject the solution offered by Copernicus. Copernicus himself, when he explained why his mind had turned to a possible new celestial system, mentioned amongst other things the divergent views that he found already in existence. The Ptolemaic system would be described as the Ptolemaic hypothesis, and we even find the Copernican theory described by one of its supporters as "the revision of the hypotheses". Many of us have gone too far perhaps in imagining a cast-iron Ptolemaic system, to the whole of which the predecessors of Copernicus were supposed to be blindly attached.

Finally, according to Dante, all the various spheres are moved by Intelligences or Spirits, which have their various grades corresponding to the degrees of nobility that exist in the physical world. Of these, the lowliest are the angels who move the sphere of the moon; for the moon is in the humblest of the heavens; she has dark spots that show her imperfections; she is associated with the servile and poor. (It is not the moon but the sun which afford the material for romantic poetry under this older system of ideas.) Through the various Intelligences operating by means of the celestial bodies, God has shaped the material world, only touching it, so to speak, through intermediaries. What He created was only inchoate matter, and this was later moulded into a world by celestial influences. Human souls, however, God creates with His own hands; and these, again, are of a special substance—they are incorruptible. Even now, long after the creation, the heavens still continue to influence the earth, however, says Dante—Venus affecting lovers, for example, by a power that comes not from the sphere but from the planet itself, a power that is actually transmitted by its rays. The Church had struggled long against the deterministic implications of astrology, and was to continue the conflict after the time of Dante, though means were already being adopted to reconcile astrology with the Christian teaching concerning free will. Dante said that the stars influence the lower dispositions of a man, but God has given all men a soul by which they can rise above such conditioning circumstances. Occasionally one meets even with the opposite view— that the stars can influence only for good, and that it is man's own evil disposition that is responsible if he turns to sin. Those who attacked astrology often took the line that the observation of the paths of the heavenly bodies was not sufficiently accurate as yet to allow of detailed predictions. Astrologers themselves, when their prophecies were found to be inaccurate, would blame the faultiness of astronomical observation rather than the defects of their own supposed science. The controversy between the supporters and the opponents of astrology could be turned into a channel, therefore, in which it became an argument in a circle. It would appear to be the case that astrology, like witch-burning, was considerably on the increase in the sixteenth and seventeenth centuries, in spite of what we say about the beginning of modern times.

In this whole picture of the universe there is more of Aristotle than of Christianity. It was the authority of Aristotle and his successors which was responsible even for those features of this teaching which might seem to us to carry something of an ecclesiastical flavour—the hierarchy of heavens, the revolving spheres, the Intelligences which moved the planets, the grading of the elements in the order of their nobility and the view that the celestial bodies were composed of an incorruptible fifth essence. Indeed, we may say it was Aristotle rather than Ptolemy who had to be overthrown in the sixteenth century, and it was Aristotle who provided the great obstruction to the Copernican theory.

The great work of Copernicus, *De Revolutionibus Orbium*, was published in 1543, though its author would appear to have been working upon it and elaborating his system since the early years of the century. It has often been pointed out that he himself was not a great observer, and his system was not the result of any passion for new observations. This passion came into astronomy later in the century, particularly with Tycho Brahé, who himself always refused to become a follower of Copernicus, and who amongst other things introduced the practice of observing planets throughout the whole of their courses, instead of just trying to pick them out when they happened to be at special points in their orbits. It was even true that Copernicus trusted too much to the observations that had been handed down by Ptolemy himself from the days of antiquity. In one of his writings he criticises a contemporary for being too sceptical concerning the accuracy of Ptolemy's observations. The later astronomer, Kepler, said that Copernicus failed to see the riches that were within his grasp, and was content to interpret Ptolemy rather than nature. It seems to have been one of his objects to find a new system which would reconcile all recorded observations, and a disciple of his has described how he would have all these before him as in a series of catalogues. It is admitted, however, that he fell into the mistake of accepting the bad observations and the good without discrimination. One modern writer has pointed out that since he was putting forward a system which claimed to account for the same phenomena as were covered by the Ptolemaic theory, he may have been wise in not laying himself open to the charge that he was doctoring Ptolemy's observations in order to fit them to his hypothesis. Through his trust in the ancient observations, however, he allowed himself to be troubled by irregularities in the sky which did not really exist; and in one or two ways he produced needless elaborations which were calculated to hinder the acceptance of his system.

If we ask, then, why he was moved to attempt a new interpretation of the skies, he tells us that he was disturbed by the differences of opinion that had already existed amongst mathematicians. There

47

is evidence that one matter of actual observation gave him some bewilderment—he was puzzled by the variations he had observed in the brightness of the planet Mars. This was the planet which during the next century caused great difficulty to astronomers and led to most remarkable developments in astronomy. Copernicus's own system was so far from answering to the phenomena in the case of Mars that Galileo in his main work on this subject praises him for clinging to his new theory though it contradicted observation—contradicted in particular what could be observed in the behaviour of Mars. It would appear that Copernicus found a still stronger stimulus to his great work in the fact that he had an obsession and was ridden by a grievance. He was dissatisfied with the Ptolemaic system for a reason which we must regard as a remarkably conservative one—he held that in a curious way it caused offence by what one can almost call a species of cheating. Ptolemy had pretended to follow the principles of Aristotle by reducing the course of the planets to combinations of uniform circular motion; but in reality it was not always uniform motion about a centre, it was sometimes only uniform if regarded as angular motion about a point that was not the centre. Ptolemy, in fact, had introduced the policy of what was called the equant, which allowed of uniform angular motion around a point which was not the centre, and a certain resentment against this type of sleight-of-hand seems to have given Copernicus a special urge to change the system. That he was in earnest in his criticism of Ptolemy's device is shown both in the system he himself produced, and in the character of certain associated ideas that gave a strong bias to his mind.

A further point has been noted sometimes. It can best be explained, perhaps, if we imagine a competent player who stares at the draught-board until a whole chain of his rival's draughts seem to stand out in his mind, plainly asking to be removed by a grand comprehensive stroke of policy. An observer of the game can often be sensible of the way in which the particular pieces glare out at one—so many black draughts which are waiting to be taken as soon as the white opponent can secure a king. It would seem to have been the case that a mind so geometrical as that of Copernicus could look at the complicated diagram of the Ptolemaic skies and see a number of the circles which cried out to be removed provided you had a king to take them with—all of them would cancel out as soon as it occurred to you to think of the earth as being in motion. For, if the ancients ignored the fact that they—the spectators of the skies —were moving, it was inevitable that to each of the heavenly bodies they should have imputed an additional, unnecessary, complicating motion—sun, planets and stars had to have a tiresome extra circle in the diagram—and this in every case would be referable to the same formula, since it corresponded to what ought to have been each time the motion of the earth. As a geometer and a mathematician Copernicus seems to have been struck by the redundancy of so many of the wheels.

Finally in this connection it is necessary to remember the way in which Copernicus rises to lyricism and almost to worship when he writes about the regal nature and the central position of the sun. He would not stand alone if he proved to have been stimulated to genuine scientific enquiry by something like mysticism or neo-platonic sentiment. He held a view which has been associated with Platonic and Pythagorean speculations to the effect that immobility was a nobler thing than movement, and this affected his attitude to both the sun and the fixed stars. Many factors, therefore, combined to stimulate his mind, and provoke him to a questioning of the ancient system of astronomy.

He had passed a number of years in Italy at one of the most brilliant periods of the Renaissance; and here he had learned something of those Platonic-Pythagorean speculations which had become fashionable, while acquiring, no doubt, some of the improved mathematics which had resulted from the further acquaintance with the achievements of antiquity. His reverence for the ancient world is illustrated in the way he always spoke of Ptolemy; and, having seen reason to be dissatisfied with the prevailing condition of things in astronomy, he tells us that he went back to study what previous writers had had to say on the whole question. Once again, as in the case of the theory of impetus, the new development in science was assisted by hints from ancient writers, and was stimulated by the differences of opinion that had already existed in antiquity. Some later medieval writers like Nicholas of Cusa had encountered the suggestion that the earth might be in motion, and had been willing to entertain the idea; but nobody had troubled to work out the details of such a scheme, and up to the time of Copernicus the heliocentric theory had never been elaborated mathematically in order to see whether it would cover and explain the observed phenomena in the competent way in which the Ptolemaic system had proved able to do so. Only the Ptolemaic theory had hitherto possessed the advantage which the modern world would prize—the merit of having been established in a concrete way, with the demonstration that it fitted the facts (on the whole) when applied to the phenomena in detail. Copernicus may have been assisted particularly by a view transmitted to the middle ages by Martianus Capella, which regarded just the two planets, Mercury and Venus, as going round the sun. These two planets, lying between the earth and the sun and always observed in close proximity to the latter, had always presented special problems when people tried to regard them as going round the earth.

Wherever he found the hint, Copernicus made it his real task to uncover the detailed workings of the skies under the new hypothesis and to elaborate the mathematics of the scheme. His own theory was only a modified form of the Ptolemaic system—assuming the same celestial machinery, but with one or two of the wheels interchanged through the transposition of the rôles of the earth and the sun. He made it a little harder for himself by trying to work all the observed movements of the planets into a more genuine system of uniform circular motion—uniform in respect of the centre of the circle, without any conjuring-tricks with equants. He had to use the old complicated system of spheres and epicycles, however, though he could claim that his hypothesis reduced the total number of wheels from eighty to thirty-four. Although some doubt has been expressed—and he himself declared the matter to be outside his concern—he appears to have believed in the actual existence of the rotating orbs—the successive crystalline heavens—and at any rate the astronomer Kepler considered this to have been the case. It was a disadvantage of his system that it was not quite heliocentric after all—the earth did. not describe an exact circle with the sun as its centre, and, in fact, all the movements of the skies were reckoned not from the sun itself, but from the centre of the earth's orbit, which came somewhat to the side. This was significant, as it infringed the old doctrine that there must be a core of hard matter somewhere on which the other things actually hinged and turned—something more than a mere mathematical point to serve as the hub of the universe.

Since the older Ptolemaic theory had approximately accounted for the phenomena, while the Copernican system itself only accounted for them approximately, much of the argument in favour of the new hypothesis was to turn on the fact of its greater economy its cleaner mathematics and its more symmetrical arrangement. Those who were unable to believe in the motion of the earth had to admit that for calculation and prediction the Copernican theory provided a simpler and shorter method. Whereas on the older view the fixed stars moved round in one direction at a speed which seemed incredible, while the planets largely turned in the opposite way and often seemed to be at sixes and sevens with the sun, it now appeared that the motion was all in the same direction—the earth and the planets, duly spaced and all in order, swung in the same sweeping way around the sun, the time of their orbits being related to their respective distances from the latter. Only thirty-four spheres or circles were needed instead of eighty, as we have seen. And by a simple daily rotation of the earth upon its axis you were saved from the necessity of making all the skies undertake a complete revolution every twenty-four hours.

On the other hand, some of what we might regard as the beautiful economy of the Copernican system only came later—for example, when some of Copernicus's own complications and encumbrances had been taken away. And if from a purely optical standpoint or from the geometer's point of view the new hypothesis was more economical, there was another sense in which it was more prodigal; because in respect of the physics of the sixteenth century it left a greater number of separate things that required separate explanation. In any case, at least some of the economy of the Copernican system is rather an optical illusion of more recent centuries. We nowadays may say that it requires smaller effort to move the earth round upon its axis than to swing the whole universe in a twenty-four hour revolution about the earth; but in the Aristotelian physics it required something colossal to shift the heavy and sluggish earth, while all the skies were made of a subtle substance that was supposed to have no weight, and they were comparatively easy to turn, since turning was concordant with their nature. Above all, if you grant Copernicus a certain advantage in respect of geometrical simplicity, the sacrifice that had to be made for the sake of this was nothing less than tremendous. You lost the whole cosmology associated with Aristotelianism—the whole intricately dovetailed system in which the nobility of the various elements and the hierarchical arrangement of these had been so beautifully interlocked. In fact, you had to throw overboard the very framework of existing science, and it was here that Copernicus clearly failed to discover a satisfactory alternative. He provided a neater geometry of the heavens, but it was one which made nonsense of the reasons and explanations that had previously been given to account for the movements in the sky. Here, as in the case of the doctrine of impetus, it was necessary to go farther still and complete the scientific revolution before one could squarely meet the criticisms to which the new hypothesis was liable. Kepler was right, therefore, when he said that Copernicus failed to see how rich he was, and erred by remaining too close to the older system of Ptolemy.

This point becomes clear when Copernicus tries to meet the objections to his hypothesis, and particularly when he has to show how the celestial machinery could possibly work, supposing his geometrical design to be correct. We are all aware that when there are two trains it is difficult to tell whether it is ours or the other one that is moving; and this purely optical relativity of motion must have long been realised, for otherwise it could never have occurred to either the ancient world or the middle ages that it was possible to discuss whether it was the earth or the sun which moved. Anybody may grant the point concerning the relativity of motion for the sake of argument, but still this does not decide for us the crucial question—it does not tell us which of the trains it is that is actually moving. In order to discuss this question at all Copernicus had to go into the further problem of the nature and cause of movement—had to move from geometry and from the problem of the mere pattern of the skies to issues which belonged rather to physics. If one asked Copernicus why the earth and the heavenly bodies moved, he would answer: Because they were spherical or because they were attached to spherical orbs. Put a sphere anywhere in space and it would naturally revolve—it would turn without needing anybody to turn it—because it was the very nature of the sphere to rotate in this way. Whereas Aristotle had made movement depend on the total nature of the celestial bodies as such, it has been pointed out that Copernicus observed with something of the geometer's eye; for in his argument the nature of the body was decided purely by the geometrical shape, and the movement merely depended upon sphericity. All bodies, furthermore, aspired to become spheres—like water forming drops—for the simple reason that the sphere represented the perfect shape. Gravity itself could belong to the sun and moon as well as to the earth—could belong to anything spherical—since it represented the tendency of all parts of a body to come together and consolidate themselves in the form of a sphere.

In fact, from a certain point of view the actual movement of the earth falls into place as almost an incidental matter in the system of Copernicus, which, viewed geometrically, as I have already said, is just the old Ptolemaic pattern of the skies with one or two of the wheels interchanged and one or two of them taken out. If one stares long at the new system, it seems to be a set of other characteristics that begin to come out into relief, and these have the effect of putting Copernicus into remarkable contrast with both the older world and modern times. Not only is Copernicus prodded and pressed into overturning the old system by a veritable obsession for uniform circular motion (the point on which he thought that Ptolemy had shuffled and faked), but in facing the two biggest problems of his system, the dynamics of it and the question of gravitation, he gives an unexpected turn to the discussion by a very similar obsession in regard to the sphere as the perfect shape. It is amazing that at this early date he had even faced these colossal issues—matters concerning which his successors went on fumbling down to the time of Galileo and even Newton. But on the first big question: What were the dynamics of the new system?—By what physical law did bodies move in the Copernican way?—his answer was neither the doctrine of impetus as such, nor the modern law of inertia, but the view that spherical bodies must turn—the earth itself according to this principle could not help turning. If you take the other great problem: Now that the earth is no longer the centre of the universe, what about the whole Aristotelian theory of gravity?—Copernicus jumps to what in one aspect is the modern view: that not only the earth but other bodies like the sun and the moon have gravity—but he ties the whole notion to the same fundamental principle—the tendency of all things to form and consolidate themselves into spheres, because the spherical shape is the perfect one. That is why his synthesis is so colossal—he not only replaces Ptolemy's astronomy but he attacks Aristotle's physics on matters of the profoundest principle. And the passion which is the motor behind everything is connected with what might seem to us almost an obsession for circularity and sphericity— one which puts the ancient Ptolemy into the shade. When you go down, so to speak, for the third time, long after you have forgotten everything else in this lecture, there will still float before your eyes that hazy vision, that fantasia of circles and spheres, which is the trade-mark of Copernicus, the very essence of undiluted Copernican thought. And though it had influence in the sixteenth century we must note that it never came in that significant way into the ideas of the seventeenth century or into the science of the modern world.

In general, it is important not to overlook the fact that the teaching of Copernicus is entangled (in a way that was customary with the older type of science) with concepts of value, teleological explanations and forms of what we should call animism. He closes an old epoch much more clearly than he opens any new one. He is himself one of those individual makers of world-systems, like Aristotle and Ptolemy, who astonish us by the power which they showed in producing a synthesis so mythical—and so irrelevant to the present day—that we should regard their work almost as a matter for æsthetic judgment alone. Once we have discovered the real character of Copernican thinking, we can hardly help recognising the fact that the genuine scientific revolution was still to come.

Within the framework of the older system of ideas, Copernicus was unable to clinch his argument. To the old objection that if the earth rotated its parts would fly away and it would whirl itself into pieces, he gave an unsatisfactory answer—he said that since rotation was for the earth a natural movement, the evil effects could not follow, for the natural movement of a body could never be one that had the effect of destroying the nature of that body. It was the argument of a man who still had one foot caught up in Aristotelianism himself, though perhaps precisely because it seems to be archaic to us it was more appropriate to those conservative people whose objections required to be met in the sixteenth century. When it was said that if the world was rushing from west to east (in the way that Copernicus suggested) the air would be left behind and there would be a continual wind from east to west, he still answered somewhat in the terms of the ancient physics, and said the air must go round with the globe because of the earthiness which the atmosphere itself contained and which rotated in sympathy with everything else that was earthy. He was not much more successful when he tried to take the argument about the earth whirling itself to pieces, and turn it against the possible critics of his theory, saying that the skies themselves, if they were to rotate so quickly as was assumed, would be broken into fragments by the operation of the very same laws. The skies and the heavenly bodies, as we have seen, were supposed to have no weight—on the Aristotelian theory they were not regarded as being subject to the operation of what we now call centrifugal force. Galileo himself apparently made mistakes when trying to meet the argument that the world would fly to pieces if it were turning on its own axis. That whole question of centrifugal force proved to be a serious obstruction to the acceptance of the Copernican system in the sixteenth century, and was only made manageable by the work of men like Huygens, whose writings appeared over a hundred years after *De Revolutionibus Orbium*. The truth was that Ptolemy in ancient times had rejected the hypothesis of the movement of the earth, not because he had failed to consider it, but because it was impossible to make such an hypothesis square with Aristotelian physics. It was not until Aristotelian physics had been overthrown in other regions altogether that the hypothesis could make any serious headway, therefore, even in modern times.

THE DOWNFALL OF ARISTOTLE AND PTOLEMY

As the crucial stage in the grand controversy concerning the Ptolemaic system does not seem to have been treated organically, and is seldom or never envisaged as a whole, it is necessary that we should put together a fairly continuous account of it, so that we may survey the transition as a whole. A bird's-eye view of the field should be of some significance for the student of the scientific revolution in general, especially as the battles come in crescendo and rise to their greatest intensity in this part of the campaign.

It would be wrong to imagine that the publication of Copernicus's great work in 1543 either shook the foundations of European thought straight away or sufficed to accomplish anything like a scientific revolution. Almost a hundred and fifty years were needed before there was achieved a satisfactory combination of ideas—a satisfactory system of the universe—which permitted an explanation of the movement of the earth and the other planets, and provided a framework for further scientific development. Short of this, it was only a generation after the death of Copernicus—only towards the close of the sixteenth century—that the period of crucial transition really opened and the conflict even became intense. And when the great perturbations occurred they were the result of very different considerations—the result of events which would have shaken the older cosmos almost as much if Copernicus had never even written his revolutionary work. Indeed, though the influence of Copernicus was as important as people generally imagine it to have been, this influence resulted not so much from the success of his actual system of the skies, but rather from the stimulus which he gave to men who in reality were producing something very different.

When Copernicus's work first appeared it provoked religious objections, especially on Biblical grounds, and since the Protestants were the party particularly inclined to what was called Bibliolatry, some scathing condemnations very soon appeared from their side— for example, from Luther and Melanchthon personally. One may suspect that unconscious prejudice had some part in this, and that the Aristotelian view of the universe had become entangled with

56

Christianity more closely than necessity dictated; for if the Old Testament talked of God establishing the earth fast, the words were capable of elastic interpretation, and Biblical exegesis in previous centuries had managed to get round worse corners than this. In any case, if the Old Testament was not Copernican, it was very far from being Ptolemaic either. And it gives something of a blow to Aristotle and his immaculate fifth essence, surely, when it says that the heavens shall grow old as a garment, and, talking of God, tells us that the stars and the very heavens themselves are not pure in His sight. The prejudice long remained with the Protestants, and when a few years ago the Cambridge History of Science Committee celebrated in the Senate House the tercentenary of the visit to England of the great Czech educator Comenius or Komensky, the numerous orations overlooked the fact that he was anti-Copernican and that his text-books, reprinted in successive editions throughout the seventeenth century, were a powerful influence in the Protestant world on the wrong side of the question. On the other hand, Copernicus was a canon in the Roman Catholic Church and high dignitaries of that Church were associated with the publication of his book. The comparatively mild reception which the new view received on this side led only recently to the enunciation of the view that the Roman Catholics, being slow in the uptake, took nearly fifty years to see that Copernicus was bound to lead to Voltaire. The truth was, however, that the question of the movement of the earth reached the stage of genuine conflict only towards the end of the sixteenth century, as I have said. By that time—and for different reasons altogether—the religious difficulties themselves were beginning to appear more serious than before.

Although Copernicus had not stated that the universe was infinite —and had declared this issue to belong rather to the province of the philosopher—he had been compelled, for a reason which we shall have to consider later, to place the fixed stars at what he called an immeasurable distance away. He was quickly interpreted—particularly by some English followers—as having put the case in favour of an infinite universe; and unless they had some non-religious objections Christians could hardly complain of this, or declare it to be impossible, without detracting from the power and glory of God. Unfortunately, however, that *enfant terrible* amongst sixteenth-century Italian speculators, Giordano Bruno, went further and talked of the actual existence of a plurality of worlds. There arose more seriously than ever before the question: Did the human beings in other worlds need redemption? Were there to be so many appearances of Christ, so many incarnations and so many atonements throughout the length and breadth of this infinite universe? That question was much more embarrassing than the purely Biblical issue which was mentioned earlier; and the unbridled speculations of Bruno, who was burned by the Inquisition for a number of heresies in 1600, were a further factor in the intensification of religious fear on the subject of the Copernican system.

Apart from all this, it is remarkable from how many sides and in how many forms one meets the thesis that is familiar also in the writings of Galileo himself—namely, the assertion that it is absurd to suppose that the whole of this new colossal universe was created by God purely for the sake of men, purely to serve the purposes of the earth. The whole outlay seemed to be too extravagant now that things were seen in their true proportions and the object had come to appear so insignificant. At this later stage the resistance to the Copernican hypothesis was common to both Roman Catholics and Protestants, though in England itself it appears to have been less strong than in most other places. The Protestant astronomer, Kepler, persecuted by the Protestant Faculty at Tübingen, actually took refuge with the Jesuits in 1596. Both the Protestant, Kepler, and the Roman Catholic, Galileo, ventured into the realms of theology by addressing their co-religionists and attempting to show them that the Copernican system was consistent with a fair interpretation of the words of Scripture. Galileo made excellent use of St. Augustine, and for a time he received more encouragement in the higher ecclesiastical circles in Rome than from his Aristotelian colleagues in the university of Padua. In the long run it was Protestantism which for semi-technical reasons had an elasticity that enabled it to make alliance with the scientific and the rationalist movements, however. That process in its turn greatly altered the character of Protestantism from the closing years of the seventeenth century, and changed it into the more liberalising movement of modern times.

The religious obstruction could hardly have mattered, however, if it had not been supported partly by scientific reasons and partly by the conservatism of the scientists themselves. It has been pointed out by one student that to a certain degree it was the astrologers who were the more ready to be open-minded on this subject in the sixteenth century. Apart from the difficulties that might be involved in the whole new synthesis which Copernicus had provided (and which, as we have seen, included a quasi-superstitious reliance upon the virtues of circles and the behaviour of spheres as such), there were particular physical objections to the attribution of movement to the earth, whether on the plan put forward by Copernicus or in any other conceivable system. Copernicus, as we have seen, had tried to meet the particular objections in detail, but it will easily be understood that his answers, which we have already noted, were not likely to put the matter beyond controversy.

Copernicus himself had been aware that his hypothesis was open to objection in a way that has not hitherto been mentioned. If the earth moved in a colossal orbit around the sun, then the fixed stars ought to show a slight change of position when observed from opposite sides of the orbit. In fact, there is a change but it is so slight that for three centuries after Copernicus it was not detected, and Copernicus had to explain what then appeared to be a discrepancy by placing the fixed stars so far away that the width of the earth's orbit was only a point in comparison with this distance. If the Ptolemaic theory strained credulity somewhat by making the fixed stars move at so great a pace in their diurnal rotation, Copernicus strained credulity in those days by what seemed a corresponding extravagance —he put the fixed stars at what men thought to be a fabulous distance away. He even robbed his system of some of its economy and its symmetry; for after all the beautiful spacing between the sun and the successive planets he found himself obliged to put a prodigal wilderness of empty space between the outermost planet, Saturn, and the fixed stars. The situation was even more paradoxical than this. When Galileo first used a telescope, one of his initial surprises was to learn that the fixed stars now appeared to be smaller than they had seemed to the naked eye; they showed themselves, he said, as mere pin-points of light. Owing to a kind of blur the fixed stars appear to be bigger than they really ought to appear to the naked eye, and Copernicus, living before that optical illusion had been clarified, was bound to be under certain misapprehensions on this subject. Even before his time some of the fixed stars had seemed to strain credulity when the attempt had been made to calculate their size on the basis of their apparent magnitude. His removal of them to a distance almost immeasurably farther away (while their apparent magnitude remained the same, of course, to the terrestrial observer) made it necessary to regard them as immensely bigger still, and strained a credulity which had been stretched over-far already.

Beyond this there was the famous objection that if the world were rushing from west to east a stone dropped from the top of a tower ought to be left behind, falling therefore well to the west of the tower. The famous Danish astronomer, Tycho Brahé, took this argument seriously, however absurd it might appear to us, and he introduced the new argument that a cannon-ball ought to carry much farther one way than the other, supposing the earth to be in motion. This argument had a novel flavour that made it particularly fashionable in the succeeding period.

In the meantime, however, certain other important things had been happening, and as a result of these it gradually became clear that great changes would have to take place in astronomy—that, indeed, the older theories were unworkable, whether the Copernican hypothesis should happen to be true or not. One of these was the appearance of a new star in 1572—an event which one historian of science appears to me to be correct in describing as a greater shock to European thought than the publication of the Copernican hypothesis itself. This star is said to have been brighter in the sky than anything except the sun, the moon and Venus—visible even in daylight sometimes—and it shone through the whole of the year 1573, only disappearing early in 1574. If it was a new star it contradicted the old view that the sublime heavens knew neither change nor generation nor corruption, and people even reminded themselves that God had ceased the work of creation on the seventh day. Attempts were made to show that the star existed only in the sublunary region, and even Galileo later thought it necessary to expose the inaccurate observations which were selected from the mass of available data to support this view. After all, Copernicus had only put forward an alternative theory of the skies which he claimed to be superior to the ancient one—now men were meeting inconvenient facts which sooner or later they would have to stop denying.

In 1577 a new comet appeared, and even some people who disbelieved the Copernican theory had to admit that it belonged to the upper skies, not to the sublunary regions—the more accurate observations which were now being made had altered the situation in regard to the observation of the whereabouts of comets. As this one cut a path straight through what were supposed to be the impenetrable crystal spheres that formed the skies, it encouraged the view that spheres did not actually exist as part of the machinery of the heavens; Tycho Brahé, conservative though he was in other respects, declared his disbelief in the reality of these orbs. In the last quarter of the sixteenth century Giordano Bruno, whom I have already mentioned, pictured the planets and stars floating in empty space, though it now became more difficult than ever to say why they moved and how they were kept in their regular paths. Also the Aristotelian theory that comets were formed out of mere exhalations from the earth, which ignited in the sphere of fire—all within the sublunary realm—was now contradicted. And those who did not wish to fly in the face of actual evidence began to modify the Aristotelian theory in detail—one man would say that the upper heavens were not unchangeable and uncorruptible; another would say that the very atmosphere extended throughout the upper skies, enabling the exhalations from the earth to rise and ignite even in the regions far above the moon. Quite apart from any attack which Copernicus had made upon the system, the foundations of the Ptolemaic universe were beginning to shake.

It is particularly towards the end of the sixteenth century that we can recognise the extraordinary intermediate situation which existed—we can see the people themselves already becoming conscious of the transitional stage which astronomical science had reached. In 1589 one writer, Magini, said that there was a great demand for a new hypothesis which would supersede the Ptolemaic one and yet not be so absurd as the Copernican. Another writer, Mæstlin, said that better observations were needed than either those of Ptolemy or those of Copernicus, and that the time had come for "the radical renovation of astronomy". People even put forward the view that one should drop all hypotheses and set out simply to assemble a collection of more accurate observations. Tycho Brahé replied to this that it was impossible to sit down just to observe without the guidance of any hypothesis at all.

Yet that radical renovation of astronomy which Mæstlin required was being carried out precisely in the closing years of the sixteenth century; and Tycho Brahé was its first leader, becoming important not for his hypotheses but precisely because of what has been called the "chaos" of observations that he left behind for his successors. We have seen that in the last quarter of the sixteenth century he achieved practically all that in fact was achieved, if not all that was possible, in the way of pre-telescopic observation. He greatly improved the instruments and the accuracy of observation. He followed the planets throughout the whole of their courses, instead of merely trying to pick them out at special points in their orbits. We have noticed also his anti-Copernican fervour, and in one respect his actual systematising was important, though his theories were not justified by events; and when he had made his observations he did not follow them up with any development of them since he was not a remarkable mathematician. He attempted, however, to establish a compromise between the Ptolemaic and the Copernican systems—some of the planets moving around the sun, but then the sun and its planetary system moving in a great sweep around the motionless earth. This is a further illustration of the intermediate and transitional character of this period, for his compromise gained a certain following, he complained later that other men pretended to be the inventors of it, and after a certain period in the seventeenth century this system secured the adhesion of those who still refused to believe in the actual movement of the earth. He was not quite so original as he imagined, and his compromise system has a history which goes back to much earlier times.

Still more significant was the fact that the chaos of data collected and recorded by Tycho Brahé came into the hands of a man who had been his assistant for a time, Johann Kepler, the pupil of the very person, Mæstlin, who had demanded a renovation of astronomy. Kepler, therefore, emerges not merely as an isolated genius, but as a product of that whole movement of renovation which was taking place at the end of the sixteenth century. He had the advantage over Tycho Brahé in that he was a great mathematician, and he could profit from considerable advances that had taken place in mathematics during the sixteenth century. There was one further factor which curiously assisted that renovation of astronomy which we are examining at the moment, and it was a factor of special importance if the world was to get rid of the crystal spheres and see the planets merely floating in empty space. An Englishman, William Gilbert, published a famous book on the magnet in 1600 and laid himself open to the gibes of Sir Francis Bacon for being one of those people so taken up with their pet subject of research that they could only see the whole universe transposed into the terms of it. Having made a spherical magnet called a *terrella*, and having found that it revolved when placed in a magnetic field, he decided that the whole earth was a magnet, that gravity was a form of magnetic attraction, and that the principles of the magnet accounted for the workings of the Copernican system as a whole. Kepler and Galileo were both influenced by this view, and with Kepler it became an integral part of his system, a basis for a doctrine of almost universal gravitation. William Gilbert provided intermediate assistance therefore—brought a gleam of light—when the Aristotelian cosmos was breaking down and the heavenly bodies would otherwise have been left drifting blindly in empty space.

With all these developments behind him, therefore, the famous Kepler in the first thirty years of the seventeenth century "reduced to order the chaos of data" left by Tycho Brahé, and added to them just the thing that was needed—mathematical genius. Like Copernicus he created another world-system which, since it did not ultimately prevail, merely remains as a strange monument of colossal intellectual power working on insufficient materials; and even more than Copernicus he was driven by a mystical semi-religious fervour —a passion to uncover the magic of mere numbers and to demonstrate the music of the spheres. In his attempt to disclose mathematical sympathies in the machinery of the skies he tried at one moment to relate the planetary orbits to geometrical figures, and at another moment to make them correspond to musical notes. He was like the child who having picked a mass of wild flowers tries to arrange them into a posy this way, and then tries another way, exploring the possible combinations and harmonies. He has to his credit a collection of discoveries and conclusions—some of them more ingenious than useful—from which we today can pick out three that have a permanent importance in the history of astronomy. Having discovered in the first place that the planets did not move at a uniform speed, he set out to find order somewhere, and came upon the law that if a line were drawn from a given planet to the sun that line would describe equal areas in equal times. At two different points in his calculations it would appear that he made mistakes, but the conclusion was happy for the two errors had the effect of cancelling one another out. Kepler realised that the pace of the planets was affected by its nearness to the sun—a point which encouraged him in his view that the planets were moved by a power actually emitted by the sun.

His achievements would have been impossible without that tremendous improvement in observation which had taken place since the time of Copernicus. He left behind him great masses of papers which help the historian of science to realise better than in the case of his predecessors his actual manner of work and the stages by which he made his discoveries. It was when working on the data left by Tycho Brahé on the subject of the movements of Mars that he found himself faced with the problem of accounting for the extraordinary anomalies in the apparent orbit of this planet. We know how with colossal expenditure of energy he tried one hypothesis after another, and threw them away, until he reached a point where he had a vague knowledge of the shape required, decided that for purposes of calculation an ellipse might give him at any rate approximate results, and then found that an ellipse was right—a conclusion which he assumed then to be true also for the other planets.

Some people have said that Kepler emancipated the world from the myth of circular motion, but this is hardly true, for from the time of the ancient Ptolemy men had realised that the planets themselves did not move in regular circles. Copernicus had been aware that certain combinations of circular motion would provide an elliptical course, and even after Kepler we find people accounting for the new elliptical path of the planets by reference to a mixture of circular movements. The obsession on the subject of circular motion was disappearing at this time, however, for other reasons, and chiefly because the existence of the hard crystal spheres was ceasing to be credible. It had been the spheres, the various inner wheels of the vast celestial machine, that had enjoyed the happiness of circular motion, while the planet, moving by the resultant effect of various compound movements, had been realised all the time to be pursuing a more irregular course. It was the circular motion of the spheres themselves that symbolised the perfection of the skies, while the planet was like the·rear lamp of a bicycle—it might be the only thing that could actually be seen from the earth, and it dodged about in an irregular manner; but just as we know that it is really the man on the bicycle who matters, though we see nothing save the red light, so the celestial orbs had formed the essential machinery of the skies, though only the planet that rode on their shoulder was actually visible. Once the crystal spheres were eliminated, the circular motion ceased to be the thing that really mattered—henceforward it was the actual path of the planet itself that fixed one's attention. It was as though the man on the bicycle had been proved not to exist, and the rear lamp, the red light, was discovered to be sailing on its own account in empty space. The world might be rid of the myth of circular motion, but it was faced with more difficult problems than ever with these lamps let loose and no bicycle to attach them to. If the skies were like this, men had to discover why they remained in any order at all—why the universe was not shattered by the senseless onrush and the uncontrollable collidings of countless billiard-balls.

Kepler believed in order and in the harmony of numbers, and it was in his attempt to fasten upon the music of the spheres that he discovered, amongst many other things, that third of his series of planetary laws which was to prove both useful and permanent—namely, the law that the squares of the period' of the orbit were proportional to the cubes of their mean distances from the sun. By this time Kepler was not the strange mystic that he had been at first—he was no longer looking for an actual music of the spheres which could be heard by God or man, or which should be loaded with mystical content. The music of the spheres was now nothing more or less to him than mathematics as such—the purely mathematical sympathies that the universe exhibited—so that what concerned him was merely to drive ahead, for ever eliciting mathematical proportions in the heavens. In fact, we may say that this worship of numerical patterns, of mathematical relations as such, took the place of the older attempt, that was still visible in Galileo, to transpose the skies into terms of circles and spheres, and become the foundation of a new kind of astronomy. It is in this particular sense that Kepler can most properly be described as having provided an improvement upon the old superstition which had hankered only after circular motion. Furthermore, by the same route, Kepler became the apostle of a mechanistic system—the first one of the seventeenth-century kind—realising that he was aspiring to turn the universe into pure clockwork, and believing that this was the highest thing he could do to glorify God. It will be necessary to glance at the Keplerian system as a whole when we come to the problem of gravitation at a later stage of the story. We must note that, of course, Kepler believed in the motion of the earth, and showed that if this supposition were accepted the movement conformed to the laws which he had discovered for the planets in general.

Besides Kepler's three planetary laws, one final addition was being made in this period to the collection of material that spelt the doom of Ptolemy and Aristotle. In 1609 Galileo, having heard of the

discovery of the telescope in Holland, created a telescope for himself, though not before an actual sample of the Dutch instrument had appeared in Venice. Instantly the sky was filled with new things and the conservative view of the heavenly bodies became more completely untenable than ever. Two items were of particular importance. First, the discovery of the satellites of Jupiter provided a picture of what might be described as a sort of miniature solar system in itself. Those who had argued that the moon obviously goes round the earth, *ergo* in a regular heaven the celestial bodies must move about the same centre, were now confronted with the fact that Jupiter had its own moons, which revolved around it, while both Jupiter and its attendants certainly moved together either around the sun as the Copernicans said, or around the earth according to the system of Ptolemy. Something besides the earth could be shown to operate therefore as the center of motions taking place in the sky. Secondly, the sunspots now became visible and if Galileo's observations of them were correct they destroyed the basis for the view that the heavens were immaculate and unchanging. Galileo set out to demonstrate that the spots were, so to speak, part of the sun, actually revolving with it, though the Aristotelians tried to argue that they were an intervening cloud, and that some of Galileo's discoveries were really the result of flaws in the lenses of his telescope. Galileo was seriously provoked by these taunts and at this point of the story of the whole controversy with the Aristotelians flared up to an unprecedented degree of intensity, not only because the situation was ripe for it, but because Galileo, goaded to scorn by university colleagues and monks, turned his attention from questions of mechanics to the larger problem of the Aristotelian issue in general. He ranged over the whole field of that controversy, bringing to it an amazing polemical imagination, which goaded the enemy in turn.

His intervention was particularly important because the point had been reached at which there was bound to be a complete impasse unless the new astronomy could be married somehow to the new science of dynamics. The Aristotelian cosmos might be jeopardised, and indeed was doomed to destruction by the recent astronomical disclosures; yet these facts did not in the least help the enquirers over the original hurdle—did not show them how to square the movement of the earth itself with the principles of Aristotelian mechanics or how to account for the motions in the sky. Copernicus had taken one course in treating the earth as virtually a celestial body in the Aristotelian sense—a perfect sphere governed by the laws which operated in the higher reaches of the skies. Galileo complemented this by taking now the opposite course—rather treating the heavenly bodies as terrestrial ones, regarding the planets as subject to the very laws which applied to balls sliding down inclined planes. There was something in all this which tended to the reduction of the whole universe to uniform physical laws, and it is clear that the world was coming to be more ready to admit such a view.

After his construction of a telescope in 1609 and the disturbing phenomena which were revealed immediately afterwards in the skies, Galileo's relations with the Peripatetics—the worshippers of Aristotle—at the university of Padua became intensely bitter. Though for a time he met with support and encouragement in high places, and even in Rome itself, the intensified controversy led to the condemnation of the Copernican hypothesis by the Congregation of the Index in 1616. This did not prevent Galileo from producing in the years 1625-29 the series of Dialogues on *The Two Principal World-Systems* which he designed to stand as his *magnum opus* and which were to lead to his condemnation. This book traversed the whole range of anti-Aristotelian argument, not merely in the realm of astronomy, but in the field of mechanics, as though seeking to codify the entire case against the adherents of the ancient system. It stands as a testimony to the fact that it was vain to attack the Aristotelian teaching merely at a single point—vain to attempt in one corner of the field to reinterpret motion by the theory of impetus as the Parisian scholastics had done—which was only like filling the gap in one jigsaw puzzle with a piece out of a different jigsaw puzzle altogether. What was needed was a large-scale change of design— the substitution of one highly dovetailed system for another—and in a sense it appeared to be the case that the whole Aristotelian synthesis had to be overturned at once. And that is why Galileo is so important; for, at the strategic moment, he took the lead in a policy of simultaneous attack on the whole front.

The work in question was written in Italian and addressed to a public somewhat wider than the realm of learning—wider than that university world which Galileo had set out to attack. Its argument was conducted much more in the language of ordinary conversation, much more in terms of general discourse, than the present-day reader would expect—the *Dialogues* themselves are remarkable for their literary skill and polemical scorn. Galileo paid little attention to Kepler's astronomical discoveries—remaining more Copernican in his general views, more content to discuss purely circular motion in the skies, than the modern reader would expect to be the case. He has been regarded as unfair because he talked only of two principal world-systems, those of Ptolemy and Copernicus, leaving the new systems of Tycho Brahé and Johann Kepler entirely out of account. In his mechanics he was a little less original than most people imagine, since, apart from the older teachers of the impetus-theory, he had had more immediate precursors, who had begun to develop the more modern views concerning the flight of projectiles, the law of inertia and the behaviour of falling bodies. He was not original when he showed that clouds and air and everything on the earth—including falling bodies—naturally moved round with the rotating earth, as part of the same mechanical system, and in their relations with one another were unaffected by the movement, so that like the objects in the cabin of a moving ship, they might appear motionless to anybody moving with them. His system of mechanics did not quite come out clear and clean, did not even quite explicitly reach the modern law of inertia, since even here he had not quite disentangled himself from obsessions concerning circular motion. It was chiefly in his mechanics, however, that Galileo made his contributions to the solution of the problem of the skies, and here he came so near to the mark that his successors had only to continue their work on the same lines and future scientists were able to read back into his writings the views which in fact were only perfected later. Galileo's kind of mechanics had a strategic place in the story, for they had to be married to the astronomy of Kepler before the new scientific order was established. And the new dynamics themselves could not be developed merely out of a study of terrestrial motion. Galileo is important because he began to develop them with reference to the behaviour of the heavenly bodies too.

At the end of everything Galileo failed to clinch his argument —he did not exactly prove the rotation of the earth—and in the resulting situation a reader either could adopt his whole way of looking at things or could reject it *in toto*—it was a question of entering into the whole realm of thought into which he had transposed the question. It was true that the genuinely scientific mind could hardly resist the case as a whole, or refuse to enter into the new way of envisaging the matter; but when Galileo's mouthpiece was charged in the *Dialogues* with having failed to prove his case—having done nothing more than explain away the ideas that made the movement of the earth seem impossible—he seemed prepared to admit that he had not demonstrated the actual movement, and at the end of Book III he brought out his secret weapon—he declared that he had an argument which would really clinch the matter. We know that Galileo attached a crucial importance to this argument, which appears in the fourth book, and, in fact, he thought of taking the title of the whole work from this particular part of it. His argument was that the tides demonstrated the movement of the earth. He made a long examination of them and said that they were caused, so to speak, by the shaking of the vessel which contained them. This seemed to contradict his former argument that everything on the earth moved with the earth, and was as unaffected by the movement as the candle in the cabin of a moving ship. It was the combination of motions, however—the daily rotation together with the annual movement, and the accompanying strains and changes of pace—which produced the jerks, he said, and therefore set the tides in motion. Nothing can better show the transitional stage of the question even now than the fact that Galileo's capital proof of the motion of the earth was a great mistake and did nothing to bring the solution of the question nearer.

Aristotelian physics were clearly breaking down, and the Ptolemaic system was split from top to bottom. But not till the time of Newton did the satisfactory alternative system appear; and though the more modern of the scientists tended to believe in the movement of the earth from this time, the general tendency from about 1630 seems to have been to adopt the compromise system of Tycho Brahé. In 1672 a writer could say that the student of the heavens had four different world-systems from which to choose, and there were men who even talked of seven. Even at this later date an enquirer could still come forward—as Galileo had done—and claim that at last he had discovered the capital argument. The long existence of this dubious, intermediate situation brings the importance of Sir Isaac Newton into still stronger relief. We can better understand also, if we cannot condone, the treatment which Galileo had to suffer from the Church for a persumption which in his dialogues on *The Two Principal World-Systems* he had certainly displayed in more ways than one.

THE HISTORY OF
THE MODERN THEORY OF GRAVITATION

It became clear to us when we were studying the work of Copernicus that the hypothesis of the daily and annual rotation of the earth presented two enormous difficulties at the start. The first was a problem in dynamics. It was the question: What power was at work to keep this heavy and sluggish earth (as well as the rest of the heavenly bodies) in motion? The second was more complicated and requires some explanation; it was the problem of gravity. On the older theory of the cosmos all heavy bodies tended to fall to the centre of the earth, because this was the centre of the universe. It did not matter if such earthy and heavy material were located for a moment on the immaculate surface of a distant star—it would still be drawn, or rather would aspire to rush, to the same universal centre, the very middle of this earth. Indeed, supposing God had created other universes besides ours and a genuine piece of earthy material found itself in one of these, it would still tend to fall to the centre of our universe, because every urge within it would make it seek to come back to its true home. Granted an earth which described a spacious orbit around the sun, however, such a globe could no longer be regarded as the centre of the universe. In that case, how could the existence of gravity be explained? For it was still true that heavy objects seemed to aspire to reach the centre of the earth.

The two problems in question became more acute when, towards the end of the sixteenth century, men began to see the untenability of the view that the planets were kept in motion and held in their proper courses by their attachment to the great crystalline orbs that formed the series of rotating skies. It became necessary to find another reason why these heavenly bodies should keep in movement yet not drift at the mercy of chance in the ocean of boundless space. These two problems were the most critical issues of the seventeenth century, and were only solved in the grand synthesis produced by Sir Isaac Newton in his *Principia* in 1687—a synthesis which represented the culmination of the scientific revolution and established

71

the basis of modern science. Though it entails a certain amount of recapitulation, we shall be drawing the threads of our whole story together if we try to mark out the chief stages in the development of this new system of the universe.

It has been suggested that Copernicus owes to Nicholas of Cusa his view that a sphere set in empty space would begin to turn without needing anything to move it. Francis Bacon said that before the problem of the heavens could be solved it would be necessary to study the question of what he called "spontaneous rotation". Galileo, who seems at times almost to have imagined gravity as an absolute —as a kind of "pull" which the universe possessed irrespective of anything in it—drew a fancy-picture of God dropping the planets vertically until they had accelerated themselves to the required speed, and then stopping the fall, turning it into circular motion at the achieved velocity—a motion which on his principle of inertia could then be presumed to continue indefinitely. Involved in the whole discussion concerning the form of the universe was the special problem of circular motion.

Copernicus was responsible for raising these great issues and, as we have already seen, he did not fail to realise the magnitude of the problems he had set. It was his view that other bodies besides the earth—the sun and the moon, for example—possessed the virtue of gravity; but he did not mean that the earth, the sun and the moon were united in a universal gravitational system or balanced against one another in a mutual harmony. He meant that any mundane object would aspire to regain contact with the earth, even if it had been carried to the surface of the moon. The sun, the moon and the earth, in fact, had their private systems, their exclusive brands, their appropriate types of gravity. For Copernicus, furthermore, gravity still remained a tendency or an aspiration in the alienated body, which rushed, so to speak, to join its mother—it was not a case of the earth exercising an actual "pull" on the estranged body. And, as we have already seen, Copernicus regarded gravity as an example of the disposition of matter to collect itself into a sphere. The Aristotelian theory had implied the converse of this—the earth became spherical because of the tendency of matter to congregate as near as possible around its centre.

In view of the principles which were emphasised in this way in the system of Copernicus, a special significance attaches to the famous book which William Gilbert published in 1600 on the subject of the magnet. This work marks, in fact, a new and important stage in the history of the whole problem which we are discussing. I have already mentioned how, according to Aristotle, four elements underlay all the forms of sublunary matter, and one of these was called "earth"—not the soil which we can take into our hands, but a more refined and sublimated substance free from the mixtures and impurities that characterise the common earth. William Gilbert, starting from this view, held that the matter on or near the surface of the globe was waste and sediment—a purely external wrapping like the skin and hair of an animal—especially as by exposure to the atmosphere and to the influence of the heavenly bodies it was peculiarly subject to debasement and to the operation of chance and change. The authentic "earth"—Aristotle's element in its pure state—was to be found below this superficial level, and formed, in fact, the bulk of the interior of this globe. Indeed, it was neither more nor less than lodestone. This world of ours was for the most part simply a colossal magnet.

The force of the magnetic attraction was the real cause of gravity, said Gilbert, and explained why the various parts of the earth could be held together. The force of the attraction exerted was always proportional to the quantity or mass of the body exerting it—the greater the mass of the lodestone, the greater the "pull" which it exercises on the related object. At the same time, this attraction was not regarded as representing a force which could operate at a distance or across a vacuum—it was produced by a subtle exhalation or effluvium, said Gilbert. And the action was a reciprocal one; the earth and the moon both attracted and repelled one another, the earth having the greater effect because it was so greatly superior in mass. If a magnet were cut in two, the surfaces where the break had been made represented opposite poles and had a hankering to join up with one another again. Magnetism seemed to represent the tendency of parts to keep together in a whole, therefore—the tendency of bodies, of material units, to maintain their integrity. Gilbert's view of gravity carried with it an attack on the idea that any mere geometrical point—the actual centre of the universe, for example—could operate as the real attraction or could stand as the goal towards which an object moved. Aristotle had said that heavy bodies were attracted to the centre of the universe. The later scholastics who adopted the impetus-theory—Albert of Saxony, for example—had developed this view and had brought out the point that in reality it was the centre of gravity of a body which aspired to reach the centre of the universe. Gilbert, on the other hand, insisted that gravity was not an action taking place between mere mathematical points, but was a characteristic of the stuff itself, a feature of the actual particles which were affected by the relationship. What was important was the tendency on the part of matter to join matter. It was the real material of the magnet that was engaged in the process, as it exercised its influence on a kindred object.

Francis Bacon was attracted by this view of gravity, and it occurred to him that, if it was true, then a body taken down a well or a mine—into the bowels of the earth—would perhaps weigh less than at the surface of the earth, since some of the attraction· exerted from below would possibly be cancelled by magnetic counter-attraction from that part of the earth which was now above. And though there were fallacies in this hypothesis, the experiment was apparently attempted more than once in the latter half of the seventeenth century; Robert Hooke, for example, stating that he tried it on Bacon's suggestion, though he failed to reach a satisfactory result. Gilbert's views on the subject of gravity took their place amongst the prevailing ideas of the seventeenth century, though they did not remain unchallenged, and it was long confessed that the question presented a mystery. Robert Boyle wrote of gravity as being possibly due to what he called "magnetical steams" of the earth. He was prepared, however, to consider an alternative hypothesis—namely, that it was due to the pressure of matter—the air itself and the æthereal substances above the air—upon any body that happened to be underneath.

William Gilbert constructed a spherical magnet called a *terrella,* and its behaviour strengthened his belief that the magnet possesses the very properties of the globe on which we live—namely, attraction, polarity, the tendency to revolve, and the habit of "taking positions in the universe according to the law of the whole"—automatically finding its proper place in relation to the rest of the cosmos. Whatever moves naturally in nature, he said, is impelled by its own force and "by a consentient compact of other bodies"; there was a correspondence between the movement of one body and another so that they formed a kind of choir; he described the planets as each observing the career of the rest and all chiming in with one another's movements. That gravitational pull towards the centre affected not merely bodies on the earth, he said, but operated similarly with the sun, the moon, etc., and these also moved in circles for magnetic reasons. Magnetism, furthermore, was responsible for the rotation of the earth and the other heavenly bodies on their axes. And it was not difficult to achieve rotation even in the case of the earth, he said, because as the earth has a natural axis it is balanced in equilibrium—its parts have weight but the earth itself has no weight—it "is set in motion easily by the slightest cause". He held that the moon always turned the same face to the earth because it was bound to the earth magnetically. But, like Copernicus, he regarded the sun as the most powerful of all the heavenly bodies. The sun, he said, was the chief inciter of action in nature.

In a curious manner the wider theories of Gilbert had found the way prepared for them and had had their prospects somewhat facilitated. Since the fourteenth century there had existed a theory that some magnetic attraction exerted by the moon was responsible for the tides. Such an idea came to be unpopular amongst the followers of Copernicus, but it appealed to astrologers because it supported the view that the heavenly bodies could exercise an influence upon the earth. In the very year after the publication of Copernicus's great treatise—that is to say, in 1544—a work was produced which attributed the tides to the movement of the earth, and Galileo, as we have already seen, was to make this point one of his capital arguments in favour of the Copernican revolution. It was in reply to Galileo that the astrologer Morin put forward a view which had already appeared earlier in the century—namely, that not only the moon but also the sun contributed to affect the tides. Galileo at one time was prepared to adopt the more general theories of Gilbert in a vague kind of way, though he did not pretend that he had understood magnetism or the mode of its operation in the universe. He regretted that Gilbert had been so much a mere experimenter and had failed to mathematise magnetic phenomena in what we have seen to be the Galileian manner.

Even earlier than Galileo, however, the great astronomer Kepler had been influenced by Gilbert's book, and it appears that he had been interested somewhat in magnetism before the work of Gilbert had been published. Kepler made important contributions to the subject, and was so far influenced by the magnetic theory that he turned the whole problem of gravity into a problem of what we call attraction. It was no longer a case of a body aspiring to reach the earth, but rather, it was the earth which was to be regarded as drawing the body into its bosom. Put a bigger earth near to this one, said Kepler, and this earth of ours would acquire weight in relation to the bigger one and tend to fall into it, as a stone falls on to the ground. And, as in Gilbert's case, it was not now a mathematical point, not the centre of the earth, that exercised the attraction, but matter itself and every particle of matter. If the earth were a sphere, the stone would tend to move towards its centre for that reason, but if the earth were differently shaped—if one of its surfaces were an irregular quadrilateral, for example—the stone would move towards different points according as it approached the earth from one side or another. Kepler further showed that the attraction between bodies was mutual—the stone attracts the earth as well as the earth a stone—and if there were nothing to interfere with the direct operation of gravity, then the earth and the moon would approach one another and meet at an intermediate point—the earth covering one-fifty-fourth of the distance (assuming it to be of the same density as the moon) because it was fifty-four times as big as the moon. It was their motion in their orbits which prevented the earth and its satellite from coming into collision with one another in this way.

In Kepler we see that curious rapprochement between gravity and magnetism which was already visible in Gilbert, whom he admired so greatly, and which is explicit in later seventeenth-century writers. As in the example of the broken magnet this gravity could be described as a tendency in cognate bodies to unite. Kepler belongs also to the line of writers who believed that the tides were caused by the magnetic action of the moon; and he has been criticised on the ground that his chains of magnetic attraction, which he pictures as streaming out of the earth, were so strong as to have made it impossible to hurl a projectile across them. He did not quite reach the idea of universal gravitation, however—for example, he did not regard the fixed stars as being terrestrial bodies by nature and as having gravity, though he knew that Jupiter threw shadows and Venus had no light on the side away from the sun. Like Bacon, he seems to have regarded the skies as becoming more æthereal—more unlike the earth—as they receded from our globe and as one approached the region of the fixed stars. Also, he regarded the sun as a special case, with, so to speak, a gravity of its own.

Having noted that the speed of planets decreased as the planet became more distant from the sun, he regarded this as a confirmation of the view to which he was mystically attached in any case—

namely, that the sun was responsible for all the motion in the heavens, though it acted by a kind of power which diminished as it operated at a farther range. He held that the planets were moved on their course by a sort of virtue which streamed out of the sun—a force which moved round as the sun itself rotated and which operated, so to speak, tangentially on the planet. He once called this force an *effluvium magneticum* and seemed to regard it as something which was transmitted along with the rays of light. If the sun did not rotate, he said, the earth could not revolve around it, and if the earth did not rotate on its axis, the moon in turn would not revolve around our globe. The rotation of the earth on its axis was largely caused by a force inherent in the earth, said Kepler, but the sun did something also to assist this movement. Granting that the earth rotates 365 times in the course of the year, he thought that the sun was responsible for five of these.

Kepler knew nothing of the modern doctrine of inertia which assumes that bodies will keep in motion until something intervenes to stop them or to deflect their course. In his theory the planets required a positive force to push them around the sky and to keep them in motion. He had to explain why the motion was elliptical instead of circular and for this purpose he made further use of magnetism—the axis of the planets, like that of the earth, always remained in one direction, and at a given angle, so that now the sun drew these bodies in, now it pushed them away, producing therefore an elliptical orbit. The force with which the planets were propelled, however, did not radiate in all directions and distribute itself indiscriminately throughout the universe like light, but moved from the sun only along the plane of the planets' orbit. The force had to know, so to speak, where to find its object, therefore—not ranging over the whole void but aiming its shafts within the limits of a given field. In a similar way, the idea that the attracting body must be sensible of its object—the earth must know where the moon was located in order to direct its "pull" to that region—was one of the obstacles to the theory of an attraction exerted by bodies on one another across empty space.

The world, then, seemed to be making a remarkable approach to the modern view of gravitation in the days of Kepler and Galileo, and many of the ingredients of the modern doctrine were already there. At this point in the story, however, an important diversion occurs, and it was to have an extremely distracting effect even long after the time of Sir Isaac Newton himself. René Descartes—who, as we have seen, had undertaken, so to speak, to reconstruct the universe, starting with only matter and motion, and working deductively—produced a world-system which it is easy for us to underestimate today, unless we remember the influence that it had even on great scientists for the rest of the century and still later. It is only in retrospect and perhaps through optical illusions that—as in the case of more ancient attempts to create world-systems—we may be tempted to feel at this point that the human mind, seeking too wide a synthesis and grasping it too quickly, may work to brilliant effect, yet only in order to produce future obstructions for itself.

We have already seen that, in spite of all his attempts to throw overboard the prejudices of the past, Descartes was liable to be misled by too easy an acceptance of data that had been handed down by scholastic writers. It is curious to note similarly that two grand Aristotelian principles helped to condition the form of the universe as he reconstructed it—first, the view that a vacuum is impossible, and secondly, the view that objects could only influence one another if they actually touched—there could be no such thing as attraction, no such thing as action at a distance. As a result of this, Descartes insisted that every fraction of space should be fully occupied all the time by continuous matter—matter which was regarded as infinitely divisible. The particles were supposed to be packed so tightly that one of them could not move without communicating the commotion to the rest. This matter formed whirlpools in the skies, and it was because the planets were caught each in its own whirlpool that they were carried round like pieces of straw—driven by the matter with which they were in actual contact—and at the same time were kept in their proper places in the sky. It was because they were all similarly caught in a larger whirlpool, which had the sun as its centre, that they (and their particular whirlpools) were carried along, across the sky, so that they described their large orbits around the sun. Gravity itself was the result of these whirlpools of invisible matter which had the effect of sucking things down towards their own centre. The mathematical principles governing the whirlpool were too difficult to allow any great precision at this time in such a picture of the machinery of the universe. The followers of Descartes laid themselves open to the charge that they reconstructed the system of things too largely by deduction and insisted on phenomena which they regarded as logically necessary but for which they could bring no actual evidence. In the time of Newton the system of Descartes and the theory of vortices or whirlpools proved to be vulnerable to both mathematical and experimental attack.

At the same time certain believers in the *plenum*—in the Cartesian idea of a space entirely filled with matter—contributed further ingredients to what was to become the Newtonian synthesis. Descartes himself achieved the modern formulation of the law of inertia—the view that motion continues in a straight line until interrupted by something—working it out by a natural deduction from his theory of the conservation of movement, his theory that the amount of motion in the universe always remains the same. It was he rather than Galileo who fully grasped this principle of inertia and formulated it in all its clarity. A contemporary of his, Roberval, first enunciated the theory of universal gravitation—applying it to matter everywhere—though he did not discover any law regarding the variation in the strength of this gravitational force as it operated at various distances. He saw a tendency throughout the whole of matter to cohere and come together; and on his view the moon would have fallen into the earth if it had not been for the thickness of the ether within the intervening space—the fact that the matter existing between the earth and the moon put up a resistance which counterbalanced the effect of gravity.

This was in 1643. It was in 1665 that the next important step was taken, when Alphonso Borelli, though he followed Kepler in the view that it needed a force emanating from the sun to push the planets around in their orbits, said that the planets would fall into the sun by the effect of gravity (which he described as a "natural instinct" in bodies) if the effect of gravity were not counterbalanced by a centrifugal tendency—a tendency to leave the curve of its orbit, like a stone seeking to leave the sling. So, though he came short in that he failed to see the planets moving by their own inertial motion and failed to understand the nature of that gravity which drew the planets towards the sun, Borelli did present the picture of the planets balanced between two opposing forces—one which tended to make them fall into the sun, and another which tended to make them fly off at a tangent. In the ancient world—in a work of Plutarch's which was familiar to Kepler, for example—the moon had been compared to a stone in a sling, in the sense that its circular motion overcame the effect of gravity. Borelli was unable

79

to carry his whole hypothesis beyond the stage of vague conjecture, however, because he failed to understand the mathematics of centrifugal force.

By this date (1665), most of the ingredients of Newton's gravitational theory were in existence, though scattered in the writings of different scientists in such a way that no man held them in combination. The modern doctrine of inertia had been put forward by Descartes and was quickly gaining acceptance, though people like Borelli, who has just been mentioned, still seemed to think that they had to provide a force actually pushing the planets along their orbit. The view that gravity was universal, operating between all bodies, had also been put forward, and on this view it became comprehensible that the sun should hold in the planets and the earth should keep the moon from flying off into space. Now, in 1665, there was the suggestion that this gravitational movement was counterbalanced by a centrifugal force—a tendency of the planets to go off at a tangent and slip out of the sling that held them. All these ideas —inertia, gravitation and centrifugal force—are matters of terrestrial mechanics—they represented precisely those points of dynamics which had to be grappled with and understood before the movements of the planets and the whole problem of the skies were settled. But if you had these on the one hand, there were the findings of astronomers on the other hand, which had to be incorporated in the final synthesis—and these included Kepler's three laws of planetary motion; the one which described the orbits as elliptical; the one which said that a line between the sun and any planet covered equal areas in equal times; and the one which said that the square of the time of the orbit was proportional to the cube of the mean distance from the sun. It had to be shown mathematically that the planets would behave in the way Kepler said they behaved, supposing their motions were governed by the mechanical laws which I have mentioned.

Huygens worked out the necessary mathematics of centrifugal force, especially the calculation of the force that was necessary to hold the stone in the sling and prevent it rushing off at a tangent. He seems to have arrived at this formula in 1659, but he only published his results in this field in 1673, as an appendix to his work on the pendulum clock. It seems, however, that it never occurred to Huygens to apply his views of circular motion and centrifugal force to the planets themselves—that is to say, to the problem of the

80

skies; and he seems to have been hampered at this point of the argument by the influence of the ideas of Descartes on the subject of the heavenly bodies. In 1669 he tried to explain gravity as the sucking effect of those whirlpools of matter with which Descartes had filled the whole of space, and he illustrated this by rotating a bowl of water and showing how heavy particles in the water moved towards the centre as the rotation slowed down. He also believed at this time that circular motion was natural and fundamental—not a thing requiring to be specially explained—and the rectilinear motion in the case of falling bodies, as with the particles in the rotating bowl of water, was, so to speak, a by-product of circular motion.

A writer on Keats has attempted to show how in the period before the production of the sonnet "On Reading Chapman's Homer" the poet had been playing in his mind—and gradually making himself at home—in what might be described as the field of its effective imagery. Now there had been an experiment in terms of astronomical discovery, now an attempt to squeeze a poetical phrase from the experience of the explorer; but one after another had misfired. The mind of the poet, however, had traversed and re-traversed the field, and in the long run apparently a certain high pressure had been generated, so that when the exalted moment came—that is to say, when Chapman's *Homer* had provided the stimulus—the happy images from those identical fields rapidly precipitated themselves in the mind of the poet. The sonnet came from the pen without effort, without any apparent preparation, but, in fact, a subterranean labour had long been taking place.

So, as the seventeenth century proceeded, the minds of men had traversed and re-traversed the fields which we have been studying, putting things together this way and that, but never quite succeeding, though a certain high pressure was clearly being generated. One man might have grasped a strategic piece in the puzzle and, in a realm which at the time hardly seemed relevant, another scientist would have seized upon another piece, but neither had quite realised that if the two were put together they would be complementary. Already the scattered parts of the problem were beginning to converge, however, and the situation had become so ripe that one youth who made a comprehensive survey of the field and possessed great elasticity of mind, could shake the pieces into the proper pattern with the help of a few intuitions. These intuitions, indeed, were to be so simple in character that, once they had been achieved, any man might well ask himself why such matters had ever given any difficulty to the world.

Already the youth in question, Isaac Newton, working independently, had discovered the required formulæ relating to centrifugal force in 1665-66, before the work of Huygens on this subject had been published. He had also discovered that the planets would move in conformity with Kepler's laws if they were drawn towards the sun by a force which varied in inverse proportion to the square of their distance from the sun—in other words, he had succeeded in giving mathematical expression to the operation of the force of gravity. On the basis of these results he compared the force required to keep a stone in a sling or the moon in its orbit with the effect of gravity (that is to say, with the behaviour of falling bodies at the earth's surface), and found that the two corresponded if one made allowance for the fact that gravity varied inversely as the square of the distance. He treated the moon as though it had been a projectile tending to rush off in a straight line but pulled into a curve by the effect of the earth's gravity; and he found that the hypotheses fitted in with the theory that the force of universal gravitation varied inversely as the square of the distance. The fall which the moon had to make (as a result of the earth's drag) every second, if it was to keep its circular path bore the requisite proportion to the descent of a body falling here at the surface of the earth. The story of Newton and the apple has at least a sort of typical validity and symbolical truth—for if it was not an apple it had to be some other terrestrial falling body that served as the basis for comparison. And the essential feat was the demonstration that when the new science of terrestrial mechanics was applied to the heavenly bodies the mathematics came out correctly. Newton, in fact, had virtually achieved the synthesis in 1665-66 while still a very young man, but he was dissatisfied with certain points in the demonstrations and put the work away for twenty years, though not abandoning it entirely.

After 1666, however, the problem interested chiefly a circle of Englishmen—not merely Newton, but also Robert Hooke, Edmond Halley and Christopher Wren. Now one of them would try the experiment suggested by Bacon to demonstrate the effect of gravity upon bodies in the bowels of the earth. Now some of them would discuss the actual path described by bodies when falling on to the earth from a very great height. Now one of them would still set out to prove that at last the annual motion of the earth could be demonstrated in a convincing manner. Men who were actually members of the Royal Society worked in ignorance of some of the things which Newton had privately achieved in the middle of the 1660s, though occasionally they called upon him for help in regard to mathematical problems of particular difficulty. Some doubt seems to have been entertained (especially amongst Englishmen) on the subject of Kepler's law relating to the elliptical orbit of the planets. Newton, however, provided a demonstration of the fact that the attraction exerted on the planets made it necessary to adopt the elliptical rather than the circular theory. Altogether the 1670s represent one of the greatest decades, if not the very climax, of the scientific revolution. Both in London and in Paris there were groups of scientific workers whose achievements were particularly remarkable.

In 1674 Robert Hooke, who afterwards claimed that he had anticipated Newton in his theory of planetary motion, produced a work entitled *An attempt to prove the annual motion of the earth*. He pointed out that, apart from the force exerted by the sun on the planets, account had to be taken of the force which all the heavenly bodies must be presumed to be exerting on one another. It would appear therefore that Hooke had at least formed an impression of the kind of situation that existed in the skies, though he could not turn the whole picture into mathematical formulæ. He certainly possessed the one thing which Huygens lacked—the right idea about the relations of the heavenly bodies and the operation of gravity in the skies; but he lacked the one thing which Huygens possessed— namely, the mathematics of curvilinear motion. It was only in 1684 that another man, Edmond Halley, produced the marriage between Hooke's system of the skies and the work of Huygens on centrifugal force. Also, though Hooke recognised that the various bodies exercised a gravitational attraction on one another, and the effect of this must diminish as the distance between the bodies increased, he had not worked out the formula for this relationship between force and distance, he had not discovered the mathematics of gravitation. Newton had shown a knowledge of the correct formula in his work in the 1660s, and a number of Englishmen, including Hooke himself, possessed it later in the '70s; that is to say, they realised that the force of gravity varied inversely as the square of the distance. The formula is supposed to have been suggested by the analogy of the

diffusion of light which even the ancient world had regarded as varying in intensity in inverse proportion to the square of the distance.

It should be noted that, whereas Kepler had seen the planets as
subject to forces which emanated from the sun, the view which
Hooke had expounded and which Newton was to develop presented
a much more complicated sky—a harmonious system in which the
heavenly bodies all contributed to govern one another in a greater
or lesser degree. The satellites of Jupiter leaned or reacted on one
another as well as influencing the planet itself, while Jupiter in turn
had a still more powerful hold upon them. The planet, however,
together with its collection of satellites, was in the grip of the sun
(upon which it exerted its own small degree of attraction), and
was also within range of the influence of neighbouring planets. As
Newton remarked later, the sun was so preponderant amongst these
bodies that the influence of the smaller ones mattered little and,
similarly, one might make small account of the influence of the moon
upon the earth. At the same time it had been noticed, especially in
England, that when Jupiter and Saturn came into closest proximity
with one another their movements showed an irregularity which
was never observed at any other point in the course of their travels.
Also the moon caused a slight alteration in the earth's orbit. By
virtue of similar perturbations in the planet Uranus in 1846 astronomers were able to deduce the existence of still another planet—
Neptune—before the planet had actually been observed. The whole
system was therefore much more complicated in the 1670s than had
been envisaged in the early part of the century—the whole sky presented a more intricate set of mathematical harmonies. It was to be
the virtue of the new theory of the skies that it explained some of
the minor anomalies, and embraced a world of interactions much
more comprehensive than anything which Kepler had envisaged.

It was in the middle of the 1680s that Isaac Newton returned to
the problem. His greatest difficulty had apparently been due to the
fact that though gravity operates, as we have seen, between all particles of matter, he had to make his calculations from one mathematical point to another—from the centre of the moon to the centre
of the earth, for example. In 1685 he was able to prove, however,
that it was mathematically correct to act upon this assumption—as
though the whole mass of the moon were concentrated at its centre,
so that the whole of its gravity could be regarded as operating from
that point. It happened furthermore that though in the middle '60s

the data upon which he worked may not have been radically wrong, still in 1684 he was able to make use of more accurate observations and calculations; for in 1672 a French expedition under Jean Picard had enabled simultaneous measurements of the altitude of Mars to be taken in Cayenne and in Paris, and the results of the expedition made it possible to secure a more accurate estimate of the sun's mean distance from the earth—which was worked out at 87 million miles, coming nearer to the modern calculation of 92 million—as well as revealing still more vividly the magnitude of the solar system. It was even possible now to have more accurate measurements of the dimensions of the earth itself. The achievements of this expedition, although they had found their way into print at an earlier date, only became widely known after a publication of 1684, and these materials were the ones employed by Newton when he made his final calculations and produced his system. In the middle '80s, therefore, there were converging reasons for his return to the problem he had been dealing with twenty years before; and this time he was satisfied with his results and demonstrations, which were completed in 1686 and communicated to the world in the *Principia* in 1687.

One of Newton's objects when he promulgated his system was to show the impossibility of that theory of vortices or whirlpools which Descartes had formulated. He showed that mathematically a whirlpool would not behave in the way that Descartes had assumed —a planet caught in a whirlpool would not act in conformity with Kepler's observations on the subject of planetary motion. Furthermore, it would not be possible for a comet to cut a straight path across the whole system, from one whirlpool to another, in the way that the theory required. In any case, if the whole of space were full of matter dense enough to carry round the planets in its whirlings, the strength of so·strong a resisting-medium would have the effect of slowing down all the movements in the universe. On the other hand, it appears that even mathematicians did not immediately grasp the meaning and the importance of the *Principia,* and many people—especially those who were under the influence of Descartes —regarded Newton as unscientific in that he brought back on to the stage two things which had been driven out as superstitious— namely, the idea of a vacuum and the idea of an influence which could operate across space between bodies that did not touch one another. His "attraction" was sometimes regarded as a lapse into the old heresies which had attributed something like occult properties to matter. Actually he denied that he had committed himself to any explanation of gravity, or to anything more than a mathematical description of the relations which had been found to exist between bodies of matter. At one moment, however, he seemed privately to favour the view that the cause of gravity was in the ether (which became less dense at or near the earth, and least dense of all at or near the sun), gravity representing the tendency of all bodies to move to the place where the ether was rarer. At another time he seemed to think that this gravitation of his represented an effect that had to be produced by God throughout the whole of space—something that made the existence of God logically necessary and rescued the universe from the over-mechanisation that Descartes had achieved. And, as we have seen, Newton believed also that certain irregular phenomena in the skies—rare combinations and conjunctures, or the passage of a comet—were liable to cause a slight derangement in the clockwork, calling for the continued intervention of God.

The great contemporaries, Huygens and Leibnitz, severely criticised the Newtonian system, and their work helped to strengthen the position of the philosophy of Descartes in Europe for many years. They attempted mechanical explanations of gravity—either imputing it to the action and pressure of subtle matter pervading the universe, or looking back to the idea of magnetism. The English in general supported Newton, while the French tended to cling to Descartes, and the result was a controversy which continued well into the eighteenth century. Both Descartes and Newton were in the first rank of geometers; but the ultimate victory of Newton has a particular significance for us in that it vindicated the alliance of geometry with the experimental method against the elaborate deductive system of Descartes. The clean and comparatively empty Newtonian skies ultimately carried the day against a Cartesian universe packed with matter and agitated with whirlpools, for the existence of which scientific observation provided no evidence.

3

Nicholas Copernicus,   ON THE REVOLUTIONS OF THE
HEAVENLY BODIES

COMMENTARIOLUS

1. What were the motivations which caused Copernicus
    to place the sun rather than the earth at the cen-
    ter of the universe? In what senses did he not go
    far enough? Why was the theory nevertheless a rev-
    olutionary one?   Why did Copernicus wait so long
    to publish his theory?

2. What were the effects of Copernicus' new theory?

3. How did Kepler solve the problems in Copernicus'
    theory?

4. Does Copernicus present his findings as a mere hy-
    pothesis? How does he distinguish between appear-
    ance and reality?

5. Why do you think Copernicus dedicated his great
    work to the Pope?

        Nicholas Copernicus (1473-1543) was a Polish
priest, astronomer, and humanist. His knowledge of
ancient Greek and Roman writings showed him that Ptolemy's geocentric view was not the only one in the ancient world and encouraged him to formulate his heliocentric theory. He was very reluctant to publish his
new theory, but initially circulated his ideas privately among friends, in treatises like the *Commentariolus*.
His great work, *On the Revolutions of the Heavenly Bodies*, was
published in 1543, shortly before his death.

DEDICATION OF THE REVOLUTIONS
OF THE HEAVENLY BODIES

BY NICOLAUS COPERNICUS (1543)

TO POPE PAUL III

I CAN easily conceive, most Holy Father, that as soon as some people learn that in this book which I have written concerning the revolutions of the heavenly bodies, I ascribe certain motions to the Earth, they will cry out at once that I and my theory should be rejected. For I am not so much in love with my conclusions as not to weigh what others will think about them, and although I know that the meditations of a philosopher are far removed from the judgment of the laity, because his endeavor is to seek out the truth in all things, so far as this is permitted by God to the human reason, I still believe that one must avoid theories altogether foreign to orthodoxy. Accordingly, when I considered in my own mind how absurd a performance it must seem to those who know that the judgment of many centuries has approved the view that the Earth remains fixed as center in the midst of the heavens, if I should, on the contrary, assert that the Earth moves; I was for a long time at a loss to know whether I should publish the commentaries which I have written in proof of its motion, or whether it were not better to follow the example of the Pythagoreans and of some others, who

Nicolaus Copernicus was born in 1473 at Thorn in West Prussia, of a Polish father and a German mother. He attended the university of Cracow and Bologna, lectured on astronomy and mathematics at Rome, and later studied medicine at Padua and canon law at Ferrara. He was appointed canon of the cathedral of Frauenburg, and in this town he died in 1543, having devoted the latter part of his life largely to astronomy.

The book which was introduced by this dedication laid the foundations of modern astronomy. At the time when it was written, the earth was believed by all to be the fixed centre of the universe; and although many of the arguments used by Copernicus were invalid and absurd, he was the first modern to put forth the heliocentric theory as "a better explanation." It remained for Kepler, Galileo, and Newton to establish the theory on firm grounds.

were accustomed to transmit the secrets of Philosophy not in writing but orally, and only to their relatives and friends, as the letter from Lysis to Hipparchus bears witness. They did this, it seems to me, not as some think, because of a certain selfish reluctance to give their views to the world, but in order that the noblest truths, worked out by the careful study of great men, should not be despised by those who are vexed at the idea of taking great pains with any forms of literature except such as would be profitable, or by those who, if they are driven to the study of Philosophy for its own sake by the admonitions and the example of others, nevertheless, on account of their stupidity, hold a place among philosophers similar to that of drones among bees. Therefore, when I considered this carefully, the contempt which I had to fear because of the novelty and apparent absurdity of my view, nearly induced me to abandon utterly the work I had begun.

My friends, however, in spite of long delay and even resistance on my part, withheld me from this decision. First among these was Nicolaus Schonberg, Cardinal of Capua, distinguished in all branches of learning. Next to him comes my very dear friend, Tidemann Giese, Bishop of Culm, a most earnest student, as he is, of sacred and, indeed, of all good learning. The latter has often urged me, at times even spurring me on with reproaches, to publish and at last bring to the light the book which had lain in my study not nine years merely, but already going on four times nine. Not a few other very eminent and scholarly men made the same request, urging that I should no longer through fear refuse to give out my work for the common benefit of students of Mathematics. They said I should find that the more absurd most men now thought this theory of mine concerning the motion of the Earth, the more admiration and gratitude it would command after they saw in the publication of my commentaries the mist of absurdity cleared away by most transparent proofs. So, influenced by these advisors and this hope, I have at length allowed my friends to publish the work, as they had long besought me to do.

But perhaps Your Holiness will not so much wonder that I have ventured to publish these studies of mine, after having taken such pains in elaborating them that I have not hesitated to commit to writing my views of the motion of the Earth, as you will be curious to hear how it occurred to me to venture, contrary to the accepted view of mathematicians, and well-nigh contrary to common sense, to form a conception of any terrestrial motion whatsoever. Therefore I would not have it unknown to Your Holiness, that the only thing which induced me to look for another way of reckoning the movements of the heavenly bodies was that I knew that mathematicians by no means agree in their investigations thereof. For, in the first place, they are so much in doubt concerning the motion of the sun and the moon, that they can not even demonstrate and prove by observation the constant length of a complete year; and in the second place, in determining the motions both of these and of the five other planets, they fail to employ consistently one set of first principles and hypotheses, but use methods of proof based only upon the apparent revolutions and motions. For some employ concentric circles only; others, eccentric circles and epicycles; and even by these means they do not completely attain the desired end. For, although those who have depended upon concentric circles have shown that certain diverse motions can be deduced from these, yet they have not succeeded thereby in laying down any sure principle, corresponding indisputably to the phenomena. These, on the other hand, who have devised systems of eccentric circles, although they seem in great part to have solved the apparent movements by calculations which by these eccentrics are made to fit, have nevertheless introduced many things which seem to contradict the first principles of the uniformity of motion. Nor have they been able to discover or calculate from these the main point, which is the shape of the world and the fixed symmetry of its parts; but their procedure has been as if someone were to collect hands, feet, a head, and other members from various places, all very fine in themselves, but not proportionate to one body, and no single one corresponding in its turn to the others, so that a monster rather than a man would be formed from them. Thus in their process of demonstration which they term a "method," they are found to have omitted something essential, or to have included something foreign and not pertaining to the matter in hand. This certainly would never have happened to them if they had followed fixed principles; for if the hypotheses they assumed were not false, all that resulted therefrom would be verified indubitably. Those things which I am saying now may be obscure, yet they will be made clearer in their proper place.

Therefore, having turned over in my mind for a long time this uncertainty of the traditional mathematical methods of calculating the motions of the celestial bodies, I began to grow disgusted that no more consistent scheme of the movements of the mechanism of the universe, set up for our benefit by that best and most law abiding Architect of all things, was agreed upon by philosophers who otherwise investigate so carefully the most minute details of this world. Wherefore I undertook the task of rereading the books of all the philosophers I could get access to, to see whether any one ever was of the opinion that the motions of the celestial bodies were other than those postulated by the men who taught mathematics in the schools. And I found first, indeed, in Cicero, that Niceta perceived that the Earth moved; and afterward in Plutarch I found that some others were of this opinion, whose words I have seen fit to quote here, that they may be accessible to all:—

"Some maintain that the Earth is stationary, but Philolaus the Pythagorean says that it revolves in a circle about the fire of the ecliptic, like the sun and moon. Heraklides of Pontus and Ekphantus the Pythagorean make the Earth move, not changing its position, however, confined in its falling and rising around its own center in the manner of a wheel."

Taking this as a starting point, I began to consider the mobility of the Earth; and although the idea seemed absurd, yet because I knew that the liberty had been granted to others before me to postulate all sorts of little circles for explaining the phenomena of the stars, I thought I also might easily be permitted to try whether by postulating some motion of the Earth, more reliable conclusions could be reached regarding the revolution of the heavenly bodies, than those of my predecessors.

And so, after postulating movements, which, farther on in the book, I ascribe to the Earth, I have found by many and long observations that if the movements of the other planets are assumed for the circular motion of the Earth and are substituted for the revolution of each star, not only do their phenomena follow logically therefrom, but the relative positions and magnitudes both of the stars and all their orbits, and of the heavens themselves, become so closely related that in none of its parts can anything be changed without causing confusion in the other parts and in the whole universe. Therefore, in the course of the work I have followed this plan: I describe in the first book all the positions of the orbits together with the movements which I ascribe to the Earth, in order that this book might contain, as it were, the general scheme of the universe. Thereafter in the remaining books, I set forth the motions of the other stars and of all their orbits together with the movement of the Earth, in order that one may see from this to what extent the movements and appearances of the other stars and their orbits can be saved, if they are transferred to the movement of the Earth. Nor do I doubt that ingenious and learned mathematicians will sustain me, if they are willing to recognize and weigh, not superficially, but with that thoroughness which Philosophy demands above all things, those matters which have been adduced by me in this work to demonstrate these theories. In order, however, that both the learned and the unlearned equally may see that I do not avoid anyone's judgment, I have preferred to dedicate these lucubrations of mine to Your Holiness rather than to any other, because, even in this remote corner of the world where I live, you are considered to be the most eminent man in dignity of rank and in love of all learning and even of mathematics, so that by your authority and judgment you can easily suppress the bites of slanderers, albeit the proverb hath it that there is no remedy for the bite of a sycophant. If perchance there shall be idle talkers, who, though they are ignorant of all mathematical sciences, nevertheless assume the right to pass judgment on these things, and if they should dare to criticise and attack this theory of mine because of some passage of Scripture which they have falsely distorted for their own purpose, I care not at all; I will even despise their judgment as foolish. For it is not unknown that Lactantius, otherwise a famous writer but a poor mathematician, speaks most childishly of the shape of the Earth when he makes fun of those who said that the Earth has the form of a sphere. It should not seem strange then to zealous students, if some such people shall ridicule us also. Mathematics are written for mathematicians, to whom, if my opinion does not deceive me, our labors will seem to contribute something to the ecclesiastical state whose chief office Your Holiness now occupies; for when not so very long ago, under Leo X, in the Lateran Council the question of revising the ecclesiastical calendar was discussed, it then remained unsettled, simply because the length of the years and months, and the motions of the sun and moon were held to have been not yet sufficiently determined. Since that time, I have given my attention to observing these more accurately, urged on by a very distinguished man, Paul, Bishop of Fossombrone, who at that time had charge of the matter. But what I may have accomplished herein I leave to the judgment of Your Holiness in particular, and to that of all other learned mathematicians; and lest I seem to Your Holiness to promise more regarding the usefulness of the work than I can perform, I now pass to the work itself.

THE *COMMENTARIOLUS*

NICHOLAS COPERNICUS

SKETCH OF HIS HYPOTHESES FOR THE HEAVENLY MOTIONS

O UR ANCESTORS assumed, I observe, a large number of celestial spheres for this reason especially, to explain the apparent motion of the planets by the principle of regularity. For they thought it altogether absurd that a heavenly body, which is a perfect sphere, should not always move uniformly. They saw that by connecting and combining regular motions in various ways they could make any body appear to move to any position.

Callippus and Eudoxus, who endeavored to solve the problem by the use of concentric spheres, were unable to account for all the planetary movements; they had to explain not merely the apparent revolutions of the planets but also the fact that these bodies appear to us sometimes to mount higher in the heavens, sometimes to descend; and this fact is incompatible with the principle of concentricity. Therefore it seemed better to employ eccentrics and epicycles, a system which most scholars finally accepted.

Yet the planetary theories of Ptolemy and most other astronomers, although consistent with the numerical data, seemed likewise to present no small difficulty. For these theories were not adequate unless certain equants were also conceived; it then appeared that a planet moved with uniform velocity neither on its deferent nor about the center of its epicycle. Hence a system of this sort seemed neither sufficiently absolute nor sufficiently pleasing to the mind.

Having become aware of these defects, I often considered whether there could perhaps be found a more reasonable arrangement of circles, from which every apparent inequality would be derived and in which everything would move uniformly about its proper center, as the rule of absolute motion requires. After I had addressed myself to this very difficult and almost insoluble problem, the suggestion at length came to me how it could be solved with fewer and much simpler constructions than were formerly used, if some assumptions (which are called axioms) were granted me. They follow in this order.

Assumptions

1. There is no one center of all the celestial circles or spheres.

2. The center of the earth is not the center of the universe, but only of gravity and of the lunar sphere.

3. All the spheres revolve about the sun as their mid-point, and therefore the sun is the center of the universe.

4. The ratio of the earth's distance from the sun to the height of the firmament is so much smaller than the ratio of the earth's radius to its distance from the sun that the distance from the earth to the sun is imperceptible in comparison with the height of the firmament.

5. Whatever motion appears in the firmament arises not from any motion of the firmament, but from the earth's motion. The earth together with its circumjacent elements performs a complete rotation on its fixed poles in a daily motion, while the firmament and highest heaven abide unchanged.

6. What appear to us as motions of the sun arise not from its motion but from the motion of the earth and our sphere, with which we revolve about the sun like any other planet. The earth has, then, more than one motion.

7. The apparent retrograde and direct motion of the planets arises not from their motion but from the earth's. The motion of the earth alone, therefore, suffices to explain so many apparent inequalities in the heavens.

Having set forth these assumptions, I shall endeavor briefly to show how uniformity of the motions can be saved in a systematic way. However, I have thought it well, for the sake of brevity, to omit from this sketch mathematical demonstrations, reserving these for my larger work. But in the explanation of the circles I shall set down here the lengths of the radii; and from these the reader who is not unacquainted with mathematics will readily perceive how closely this arrangement of circles agrees with the numerical data and observations.

Accordingly, let no one suppose that I have gratuitously asserted, with the Pythagoreans, the motion of the earth; strong proof will be found in my exposition of the circles. For the principal arguments by which the natural philosophers attempt to establish the immobility of the earth rest for the most part on the appearances; it is particularly such arguments that collapse here, since I treat the earth's immobility as due to an appearance.

The Order of the Spheres

The celestial spheres are arranged in the following order. The highest is the immovable sphere of the fixed stars, which contains and gives position to all things. Beneath it is Saturn, which Jupiter follows, then Mars. Below Mars is the sphere on which we revolve; then Venus; last is Mercury. The lunar sphere revolves about the center of the earth and moves with

the earth like an epicycle. In the same order also, one planet surpasses another in speed of revolution, according as they trace greater or smaller circles. Thus Saturn completes its revolution in thirty years, Jupiter in twelve, Mars in two and one-half, and the earth in one year; Venus in nine months, Mercury in three.

The Apparent Motions of the Sun

The earth has three motions. First, it revolves annually in a great circle[8] about the sun in the order of the signs, always describing equal arcs in equal times; the distance from the center of the circle to the center of the sun is ½₅ of the radius of the circle.[9] The radius is assumed to have a length imperceptible in comparison with the height of the firmament;[10] consequently the sun appears to revolve with this motion, as if the earth lay in the center of the universe. However, this appearance is caused by the motion not of the sun but of the earth, so that, for example, when the earth is in the sign of Capricornus, the sun is seen diametrically opposite in Cancer, and so on. On account of the previously mentioned distance of the sun from the center of the circle, this apparent motion of the sun is not uniform, the maximum inequality being 2⅙°.[11]

[8] This great circle is the *orbis magnus* discussed above (p. 16).

[9] Here Copernicus accepts Ptolemy's view that the eccentricity was fixed (HI, 233.11-16). However, Ptolemy had put the eccentricity at ½₄ (HI, 236.19-21). Hence we may say that in the *Commentariolus* Copernicus retains a fixed eccentricity, but offers an improved determination of it. On the other hand, in *De rev.* he finds that the eccentricity is ⅓₁ (Th 211.23-25; cf. p. 160, below). Consequently he there abandons the idea of a fixed eccentricity (Th 209.27-210.1), and holds that it varies between a maximum of ½₄ and a minimum of ⅓₁ (Th 219.31-220.6, 209.11-13, 211.18).

[10] See Assumption 4, above.

[11] Let the apparent motion of the sun (or real motion of the earth) take place on the great circle (*orbis magnus*) AEP (Fig. 20). Let the motion be uniform

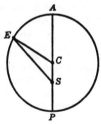

FIGURE 20

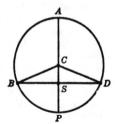

FIGURE 21

with respect to the center at C. Let the sun be at S. Let the apogee be at A, and the perigee at P. Assume that the earth starts from A and has reached any point E on the circumference. Then the line of sight ES will give the observed place of the sun, and ∠ ASE will measure the observed motion. But ∠ ACE will measure

97

The line drawn from the sun through the center of the circle is invariably directed toward a point of the firmament about 10° west of the more brilliant of the two bright stars in the head of Gemini;[12] therefore when the earth is opposite this point, and the center of the circle lies between them, the sun is seen at its greatest distance from the earth.[13] In this circle, then,

the uniform or mean motion. Now the inequality to which Copernicus refers is the difference between the uniform and the observed motions; and it is measured by ∠ CES. It is evident that when the earth (or the observed place of the sun) is at A or P, the inequality is zero.

If we draw BD ⊥ ACSP at S (Fig. 21), the inequality attains its maximum at B and D (Th 207.15-208.7; cf. HI, 220.12-16, 221.9-223.3). It is obvious that the smaller the eccentricity CS is, the smaller the maximum inequality will be. Now Ptolemy had put the maximum inequality at 2° 23′ (HI, 238.22-239.1), corresponding to an eccentricity of ¼₄ (CS:AC=1:24). Since in the *Commentariolus* Copernicus reduces the eccentricity to ¼₅, he diminishes the maximum inequality to 2° 10. And in *De rev.*, where he further reduces the eccentricity to ⅓₁, the maximum inequality is 1° 51′ (Th 212.14-16).

When he states in the *Commentariolus* that the maximum inequality, corresponding to an eccentricity of ¼₅, is 2⅙° (*duobus gradibus et sextante unius*), he is evidently writing a convenient fraction. For ¼₅ × 100,000 = 4,000; and by his Table of Chords, 4,000 subtends 2° 17½′ (Th 44.20-21). For the equivalence of Copernicus's Table of Chords with a modern table of sines see Armitage, *Copernicus*, pp. 171-73.

[12] With what fixed star are we to identify "the more brilliant of the two bright stars in the head of Gemini" (*stella lucida quae est in capite Gemelli splendidior*)? Both Gemini 1 (Castor, α Geminorum) and Gemini 2 (Pollux, β Geminorum) were described as being in the head of Gemini. They were usually differentiated as western and eastern (HII, 20.17-18, 21; 92.3-4; Th 132.29-32); thus in a later section of the *Commentariolus*, where Copernicus refers to "the star which is described as being in the head of the eastern of the two Gemini" (*stellam quae in capite Geminorum orientalis dicitur*), he is speaking of Pollux (p. 78). But in the present passage he does not employ the customary designation, and he relies on *splendidior* to indicate whether he is referring to Castor or to Pollux. Now, in the catalogues these stars were both listed as of the the second magnitude; and it therefore seems impossible to decide the question by appealing to a difference in brilliancy. However, Pollux is distinguished by its color; and it is perhaps possible that Copernicus is using *splendidior* as a color term. On this uncertain basis let us tentatively identify the star of our text with Pollux.

[13] If the preceding note is correct, then Copernicus is here locating the solar apogee about 10° west of Pollux. Now Ptolemy had put the longitude of the solar apogee at 65° 30′ (HI, 237.9-11) and the longitude of Pollux at 86° 40′ (HII, 93.4); hence the apogee was 21° 10′ west of Pollux. He held that the

the earth revolves together with whatever else is included within the lunar sphere.

The second motion, which is peculiar to the earth, is the daily rotation on the poles in the order of the signs, that is, from west to east. On account of this rotation the entire universe appears to revolve with enormous speed. Thus does the earth rotate together with its circumjacent waters and encircling atmosphere.[14]

The third is the motion in declination. For the axis of the daily rotation is not parallel to the axis of the great circle, but is inclined to it at an angle that intercepts a portion of a circum-

apogee was fixed in relation to the vernal equinox (HI, 232.18-233.16), that the equinoctial and solstitial points were constant (HI, 192.12-22), and that the fixed stars moved eastward 1° in 100 years (HII, 15.15-17). Had Ptolemy been right, the apogee should have been found, at the time of Copernicus, about 35° west of Pollux.

Copernicus reverses Ptolemy's explanation of precession; for he regards the fixed stars as constant (Assumption 5) and attributes the precessional motion to the equinoctial and solstitial points (p. 67, below). Hence, when in the present passage he asserts that the solar apogee is fixed, he means that it is fixed in relation to the fixed stars; but its distance from the vernal equinox increases, because the equinoctial points move steadily westward.

Furthermore, so long as the vernal equinox was regarded as constant, it had served as the point from which celestial longitude was measured (HII, 36.16-17). Hence if Copernicus is to utilize without error the work of the ancient astronomers, he must first reconstruct the entire history of precession. In the next section of the *Commentariolus* he lays down two of the main propositions. But it is evident that he has not yet completely formulated the theory which is outlined briefly in the *Letter against Werner* (pp. 99-101, below) and expounded fully in *De rev.* (Th 157-173; cf. pp. 111-17, below). Moreover, since celestial longitude can no longer be reckoned from the vernal equinox, some fixed star must be selected in its stead. But Copernicus has apparently not yet made a choice; throughout the *Commentariolus*, when he states a celestial position, he gives it in terms of neighboring stars, never in terms of longitude reckoned from a fixed origin. Because he does not give us sufficient data to make the correction for precession, we cannot say with precision what he then believed the longitude of the solar apogee to be.

In *De rev.* he chooses a fixed star from which to measure longitude (Th 114. 22-33, 130.6-7); he determines the longitude of the solar apogee as 96° 40' (Th 211.20-21, 25-26); and he no longer regards the doctrine of a fixed apogee as tenable: "There now emerges the more difficult problem of the motion of the solar apse . . . which Ptolemy thought was fixed . . ." (Th 216.3-4).

[14] Cf. above, p. 58, n. 4.

ference, in our time about 23½°.[15] Therefore, while the center
of the earth always remains in the plane of the ecliptic, that is,
in the circumference of the great circle, the poles of the earth
rotate, both of them describing small circles about centers equidistant from the axis of the great circle.[16] The period of this
motion is not quite a year and is nearly equal to the annual
revolution on the great circle. But the axis of the great circle
is invariably directed toward the points of the firmament which
are called the poles of the ecliptic. In like manner the motion
in declination, combined with the annual motion in their joint
effect upon the poles of the daily rotation, would keep these
poles constantly fixed at the same points of the heavens, if the
periods of both motions were exactly equal.[17] Now with the
long passage of time it has become clear that this inclination
of the earth to the firmament changes. Hence it is the common
opinion that the firmament has several motions in conformity
with a law not yet sufficiently understood. But the motion of

[15] In *De rev.* Copernicus states that he and certain of his contemporaries have
found this angle (which is equal to the obliquity of the ecliptic) to be not
greater than 23° 29′ (Th 76.29-77.1); and again that "in our times it is found
to be not greater than 23° 28½′" (Th 162.24-25). Newcomb's determination
of the obliquity for 1900 was 23° 27′ 8″.26; on the basis of an annual diminution
of 0″.4684, the value for 1540 would be 23° 29′ 57″ (*American Ephemeris and
Nautical Almanac for 1940*, Washington, D. C., 1938, p. xx).

[16] Müller's version is faulty. He translated: "beschreiben die beiden Pole der
Erdachse bei stets gleichbleibendem Abstand kleine Kreise um die Pole der Ekliptik" (the two poles of the earth's axis, always maintaining an equal distance,
describe small circles about the poles of the ecliptic; ZE, XII, 367). But Copernicus says plainly enough that it is the centers of the small circles that are equidistant from the poles of the ecliptic: *circulos utrobique parvos describentes in
centris ab axe orbis magni aequidistantibus;* hence the poles of the earth are not,
as Müller thought, equidistant from the poles of the ecliptic. This blunder led
Müller into another error, as we shall see below (p. 73, n. 45).

[17] This obviously requires the direction of the motion in declination to be
opposite to the direction of the annual motion. The explicit statement appears in
De rev. (where the annual revolution of the earth about the sun is termed "the
annual motion of the center" or more briefly "the motion of the center"): "Then
there follows the third motion of the earth, the motion in declination, which is
also an annual revolution but which takes place in precedence, that is, in the
direction opposite to that of the motion of the center. Since the two motions are
nearly equal in period and opposite in direction. . ." (Th 31.22-25; cf. p. 148,
below).

the earth can explain all these changes in a less surprising way. I am not concerned to state what the path of the poles is. I am aware that, in lesser matters, a magnetized iron needle always points in the same direction. It has nevertheless seemed a better view to ascribe the changes to a sphere, whose motion governs the movements of the poles. This sphere must doubtless be sublunar.

Equal Motion Should Be Measured Not by the Equinoxes but by the Fixed Stars

Since the equinoxes and the other cardinal points of the universe shift considerably, whoever attempts to derive from them the equal length of the annual revolution necessarily falls into error.[18] Different determinations of this length were made in different ages on the basis of many observations. Hipparchus computed it as 365¼ days, and Albategnius the Chaldean as $365^d \; 5^h \; 46^m$,[19] that is, $13\frac{3}{5}^m$ or $13\frac{1}{8}^m$ less than Ptolemy.[20] Hispalensis increased Albategnius's estimate by the

[18] This assertion is directed against the Ptolemaic doctrine that the length of the year must be measured by the solstices and equinoxes (HI, 192.12-22; cf. Th 309.4-9).

[19] In *De rev.* Copernicus cites Albategnius's estimate more fully as $365^d5^h46^m24^s$ (Th 193.7-8). It is to this value that he adds $13\frac{3}{5}^m$ ($= 13^m36^s$), in order to obtain the sum $365\frac{1}{4}^d$ ($= 365^d6^h$). For Albategnius's determination see C. A. Nallino, *Al-Battānī sive Albatenii opus astronomicum* (*Pubblicazioni del Reale osservatorio di Brera in Milano*, No. 40, 1899–1907), Pt. I, 42.17. It seems clear that Copernicus did not draw from a single source the historical statements made in this section. But it is altogether likely that they were in large part based upon the *Epitome in Almagestum Ptolemaei* (Venice, 1496), begun by George Peurbach and completed by Regiomontanus (for Rheticus's use of this work, see below, p. 117, n. 35). For the *Epitome* (Bk. III, Prop. 2) gave Albategnius's determination as $365^d5^h46^m24^s$ or $13\frac{3}{5}^m$ less than $365\frac{1}{4}^d$.

[20] When Copernicus wrote the *Commentariolus*, he was misinformed about the value accepted by Hipparchus and Ptolemy, for he put it at 365¼ days. But in *De rev.* he correctly states that they found the year less than 365¼ days by $\frac{1}{300}$th of a day, or $365^d5^h55^m12^s$ (Th 191.31-192.3, 192.21-23, 237.13-15; HI, 207.24-208.14). The *Epitome* (*loc. cit.*) cited Hipparchus's determination as $365\frac{1}{4}^d$, but quoted Ptolemy's value correctly. It should be noted that a work contemporary with the *Commentariolus* states: "Hipparchus thought that the year consisted of 365¼ days. Although he says that it was a fraction less than the complete quarter, he ignored the fraction, since he judged it to be imperceptible" (Augustinus Ricius, *De motu octavae sphaerae*, Trino, 1513, fol. e6r; Paris, 1521, p. 40 r).

20th part of an hour, since he determined the tropical year as
$365^d\ 5^h\ 49^m$.[21]

[21] Prowe (PII, 191 n) and Müller (ZE, XII, 368, n. 41; the reference to
De rev. should be III, xiii, not III, liii) followed Curtze (MCV, I, 10 n) in
supposing that Hispalensis, i.e., from Hispalis = Seville, here means Isidore of
Seville. In Copernicus's view precession attained its greatest rapidity in the time
of Albategnius; thereafter diminution set in: "From these computations it is
clear that in the 400 years before Ptolemy the precession of the equinoxes was
less rapid than in the period from Ptolemy to Albategnius, and that in this
same period it was more rapid than in the interval from Albategnius to our times"
(Th 162.14-17; cf. p. 113, below). Therefore the shortest length of the tropical
year fell in the time of Albategnius; and the increase noted by Hispalensis must
be associated with a later astronomer. This chronological consideration rules out
Isidore immediately. Moreover, an examination of the astronomical portions of
his extant works (J. P. Migne, *Patrologia Latina*, Vols. LXXXI–LXXXIV) shows
that he gives 365 days as the length of both the tropical and sidereal years.

Who, then, is Hispalensis? Jābir ibn Aflaḥ? In 1534 Peter Apian's *Instru-
mentum primi mobilis* was published together with Gebri filii Affla Hispalensis
. . . *Libri IX de astronomia.* A copy was given by Rheticus to Copernicus (MCV,
I, 36), and hence it did not get into his hands before 1539 (PII, 377.11-12).
But all our evidence points to 1533 as the very latest year in which the *Commentariolus* could have been written. Moreover, Jābir (*op. cit.*, pp. 38-39)
simply repeats the Hipparchus-Ptolemy estimate of the length of the tropical
year. Clearly he is not the Hispalensis to whom Copernicus refers.

In his *Stromata Copernicana* (Cracow, 1924), p. 353, Birkenmajer correctly identified
Copernicus' "Hispalensis" with Alfonso de Cordoba Hispalensis. The latter, who
usually called himself *Alfonsus artium et medicinae doctor,* corrected Abraham Zacuto's
*Almanach perpetuum exactissime nuper emendatum omnium celi motuum cum additionibus
in eo factis tenens complementum* (Venice, 1502). On fol. a2r a letter is addressed to
him as *Alfonso hispalensi de corduba artium et medicinae doctori.* His correction of
Zacuto's *Almanach perpetuum* was published by Peter Liechtenstein at Venice on
July 15, 1502, while Copernicus was a student at the nearby University of Padua.
Alfonso Hispalensis' statement concerning the length of the year occurs on fol. a1v,
where he corrects a computation of Zacuto and says: . . . *dividas per numerum
dierum anni* .365. *et quartam minus undecim minutis hore* . . . (divide by the number
of days in a year, 365¼ minus eleven minutes = $365^d5^h49^m$). This direct statement
was overlooked by Birkenmajer, who thought he found nearly the same length of
the tropical year by implication in the tables (which, however, were due to Zacuto
and not to Alfonso Hispalensis). Birkenmajer also misread the second word in the
volume's title, where "perpetuu3" = "perpetuum," not "perpetuum et" (Adriano
Cappelli, *Lexicon abbreviaturarum,* 5th ed., Milan, 1954, p. XXXII). The *Almanach
perpetuum* belonging to the library of the Ermland cathedral chapter (ZE, V, 375)
may or may not have been a copy of the Venice, 1502 edition. The copy of that edition
in the library of Upsala University (Pehr Fabian Aurivillius, *Catalogus librorum
impressorum bibliothecae r. academiae Upsaliensis,* Upsala, 1814, p. 1002) lacks the
page on which the entry *Liber capit. Varm.* would have appeared, had the volume

Lest these differences should seem to have arisen from errors of observation, let me say that if anyone will study the details carefully, he will find that the discrepancy has always corresponded to the motion of the equinoxes. For when the cardinal points moved 1° in 100 years, as they were found to be moving in the age of Ptolemy,[22] the length of the year was then what Ptolemy stated it to be. When however in the following centuries they moved with greater rapidity, being opposed to lesser motions, the year became shorter; and this decrease corresponded to the increase in precession. For the annual motion was completed in a shorter time on account of the more rapid recurrence of the equinoxes. Therefore the derivation of the equal length of the year from the fixed stars is more accurate. I used Spica Virginis[23] and found that the year has always been 365 days, 6 hours, and about 10 minutes,[24] which is also the estimate of the ancient Egyptians.[25] The same method must be employed also with the other motions of the

---

once belonged to the library of the Ermland chapter (Birkenmajer, *Stromata*, p. 300).

Since "Hispalensis" in the *Commentariolus* means the *Almanach perpetuum* of 1502, it follows that Copernicus wrote the *Commentariolus* after July 15 of that year. If the entry . . . *sexternus Theorice asserentis Terram moveri, Solem vero quiescere* . . . (a manuscript of six leaves expounding the theory of an author who asserts that the earth moves while the sun stands still) in the catalogue of his books drawn up on May 1, 1514, by Matthew of Miechow (1457–1523), professor at the university of Cracow, refers to the *Commentariolus*, then its date of composition is narrowed down to the dozen years between July 15, 1502 and May 1, 1514.

[22] HII, 15.6–16.2.   [23] Virgo 14 (HII, 102.16; Th 136.10), α Virginis.

[24] Copernicus's estimate of the length of the sidereal year is stated more exactly in *De rev.* as $365^d6^h9^m40^s$ (Th 195.29–196.2); Curtze misquotes the estimate as $365^d6^h8^m40^s$ (MCV, I, 10 n), and Prowe repeats the misstatement (PII, 191 n). Newcomb's determination (1900) is $365^d.25636042 = 365^d6^h9^m9^s.54$ (*American Ephemeris for 1940*, p. xx).

[25] Copernicus apparently derived this information from the *Epitome*. It stated (*loc. cit.*) that the value found by the ancient Egyptians was $365\frac{1}{4}^d + \frac{1}{130}^d$ ($= 365^d6^h11^m$). The Latin translation of Albategnius, which was printed at Nuremberg in 1537, likewise ascribed to certain ancient Egyptian and Babylonian astronomers a year consisting of $365\frac{1}{4}^d + \frac{1}{131}^d = 365^d6^h11^m$ (Nallino, *Al-Battānī*, I, 40.28–29, 204–9; cf. below, p. 117, n. 34). So far as I am aware, no determination of the length of the year more precise than $365\frac{1}{4}^d$ has been discovered among the papyri or other documents surviving from ancient Egypt.

planets, as is shown by their apsides, by the fixed laws of their motion in the firmament, and by heaven itself with true testimony.

The Moon

The moon seems to me to have four motions in addition to the annual revolution which has been mentioned. For it revolves once a month on its deferent circle about the center of the earth in the order of the signs.[26] The deferent carries the epicycle which is commonly called the epicycle of the first inequality or argument, but which I call the first or greater epicycle.[27] In the upper portion of its circumference this greater epicycle revolves in the direction opposite to that of the deferent,[28] and its period is a little more than a month. Attached

[26] The loss of a leaf from V creates a lacuna which begins at this point and ends near the close of the present section. For the intervening text we must rely on S alone.

[27] The meaning of *anni* is not clear to me, and I have omitted it from the translation. Müller rendered the passage as follows: "wir nennen ihn einfach den ersten, den Haupt- oder Jahres-Epicykel" (but which I call the first, the chief, or annual epicycle; ZE, XII, 370). There are three objections to Müller's version of *anni*. It is syntactically unsound; in Copernicus's system the first lunar epicycle has no connection with the year; Copernicus regularly employs in his lunar theory the terms "first epicycle" and "greater epicycle," but never "annual epicycle" or "epicycle of the year" (cf. Th 235.14-15, 257.7-8, 262.26, 277.22, 288.23).

[28] When the motion of a circle, in the upper portion of its circumference, is in precedence, i.e., from east to west, in the lower portion it is in consequence, from west to east; and vice versa. "Now let *abc* (Fig. 22) be the epicycle . . .

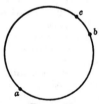

FIGURE 22

and let the motion of the epicycle be understood to be from *c* to *b* and from *b* to *a*, that is, in precedence in the upper portion and in consequence in the lower portion" (Th 251.26–252.1; cf. also Th 323.26-28, 325.21-23; PII, 349.14-16). When the direction of a motion is stated without reference to the portion of the circumference, it is the upper circumference that is understood.

to it is a second epicycle. The moon, finally, moving with this second epicycle, completes two revolutions a month in the direction opposite to that of the greater epicycle, so that whenever the center of the greater epicycle crosses the line drawn from the center of the great circle through the center of the earth (I call this line the diameter of the great circle), the moon is nearest to the center of the greater epicycle. This occurs at new and full moon; but contrariwise at the quadratures, midway between new and full moon,[29] the moon is most remote from the center of the greater epicycle. The length of the radius[30] of the greater epicycle is to the radius of the def-

Müller was evidently unfamiliar with this usage, for he detached *in superiore quidem portione* from *contra motum orbis reflexus*. He translated: "dabei führt er auf seiner Aussenseite einen ferneren Epickel mit sich" (as the first epicycle revolves, it carries with it on its surface another epicycle; ZE, XII, 370). But Rheticus explicitly states: "As the first epicycle revolves uniformly about its own center, in its upper circumference it carries the center of the small second epicycle in precedence, in its lower circumference, in consequence" (p. 134, below).

[29] Here, too, Müller blundered. For he translated *in quadraturis mediantibus iisdem* by: "zur Zeit der mittleren Quadraturen" (at the time of the mean quadratures; ZE, XII, 371). This version ignores *iisdem* and mistakes *mediare* (to halve) for *medius* (the technical astronomical term for "mean"). But Copernicus has not yet begun to discuss the lunar inequalities; all that he is stating here is the elementary fact (see p. 47, above) that the quadratures are midway between new and full moon (*iisdem*).

[30] Although *diametri*, the reading of S, cannot be checked on account of the lacuna in V, it is certainly wrong and must be changed to *semidiametri*. Computational support for this emendation is adduced in n. 32. Additional support comes from a calculation jotted down by Copernicus in his copy of the Tables of Regiomontanus (see Curtze in *Zeitschrift für Mathematik und Physik*, XIX(1874), 454-56). The note reads: *Semidiametrus orbis lunae ad epicyclium a* $\frac{10}{1\frac{1}{18}}$; *epicyclus a ad b* $\frac{19}{4}$ (PII, 211); "Radius of deferent of moon to first epicycle 10:1⅟18; first epicycle to second epicycle 19:4." Throughout this series of calculations Copernicus is comparing radius with radius, never diameter with radius.

While the note was properly used by Curtze to emend another false reading (*parte* for *quarta*) in this same sentence of S, he overlooked *diametri*. Curiously enough, in citing Curtze's work Prowe speaks of the note as containing "values calculated by Copernicus for the radii of the planetary epicycles" (PII, 193 n); yet he too failed to notice the discrepancy. Had Müller compared his computations (ZE, XII, 372, n. 51) for the *Commentariolus* with the lunar numerical ratios in *De rev.*, he would surely have caught the copyist's error. It should be observed that Rheticus compares diameter with diameter when he gives the ratio of the lunar epicycles (p. 134, below).

erent as $1\frac{1}{18}:10;$ [31] and to the radius of the smaller epicycle as $4\frac{3}{4}:1.$ [32]

By reason of these arrangements the moon appears, at times rapidly, at times slowly, to descend and ascend; and to this first inequality the motion of the smaller epicycle adds two irregularities.[33] For it withdraws the moon from uniform motion on the circumference of the greater epicycle, the maximum inequality being $12\frac{1}{4}°$ of a circumference of corresponding size or diameter;[34] and it brings the center of the greater epicycle

[31] I have adopted this form for the sake of clarity and compactness. What Copernicus actually wrote may be literally translated as follows: "The length of the radius of the greater epicycle contains a tenth part of the radius of the deferent plus one-eighteenth of such tenth part." This ratio may be numerically represented by the expression $\frac{1}{10} + \frac{1}{18} \cdot \frac{1}{10} : 1$ or $1\frac{1}{18} : 10$.

[32] Literally: "(The length of the radius of the greater epicycle) contains the radius of the smaller epicycle five times minus one-fourth of the smaller radius." While Copernicus incorporated in *De rev.* the lunar theory sketched in this section, he altered the numerical components slightly (Th 258.10-11). The ratio of first epicycle to deferent is given here as $1\frac{1}{18} : 10$, which may be written $1055:10,000$; in *De rev.* it has been changed to $1097:10,000$, which may be written $1\frac{1}{10} : 10$. The ratio of first epicycle to second epicycle appears there as $1097:237$, which may be written $4.63:1$; it is given above as $4.75:1$.

[33] Although the meaning of the passage is clear, the text is faulty and simply does not parse. We might have expected *et primae quidem diversitati dupliciter variationem motus epicycli minoris ingerit* (cf. Th 257.20-21). The distance from the moon to the center of the earth varies, because the moon's orbit around the earth is really an ellipse; and the rate of the moon's apparent motion varies for the same reason. Copernicus uses the term "first inequality" to denote the variation in the moon's distance from the center of the earth and employs the first epicycle to account for it. Both the term and the geometrical device were traditional (cf. HI, 300.16-301.1).

[34] The inequality is measured by an arc of the greater epicycle, or of a circle of equal dimensions. Let AB be the greater epicycle with center at C (Fig. 23). Choose any point E on the circumference, and with E as center describe the second epicycle. Draw CM and CL tangent to the second epicycle. When the moon is at M or L, the inequality attains its maximum. Now in the *Commentariolus* CE:EM $= 4.75:1 = 100,000:$ $21,053$. Then by the Table of Chords $\angle$ ECM, which measures the maximum inequality, $= 12° 9'$ (Th 45.19-20). Hence the reading of S, *17 gradus et quadrantem*, is certainly wrong, and must be corrected to *12 gradus et quadrantem*. As in the case of the solar inequality (see above, p. 62, n. 11), Copernicus is writing a convenient fraction.

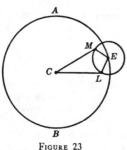

FIGURE 23

at times nearer the moon, at times further from it, within the limits of the radius of the smaller epicycle.[35] Therefore, since the moon describes unequal circles about the center of the greater epicycle, the first inequality varies considerably. In conjunctions and oppositions to the sun its greatest value does not exceed 4°56', but in the quadratures it increases to 6°36'.[36] Those who employ an eccentric circle to account for this variation[37] improperly treat the motion on the eccentric as unequal,[38]

While the reading of S cannot be checked on account of the lacuna in V, the proposed emendation is confirmed by a comparison with *De rev.* We saw above (n. 32) that in the later work Copernicus diminished the ratio CE:EM, making it 1097:237 = 4.63:1 = 100,000:21,604. It is obvious that since CE has been shortened in relation to EM, ∠ECM must increase; by the Table of Chords, it is 12° 28' (Th 45.21-22, 258.32–259.4, 264.31). Hence any such value as 17¼° for the maximum inequality in the *Commentariolus* must be rejected as a copyist's error.

[35] Reading *cum* for *eum* (PII, 193.11).

[36] The difference between the maximum in the quadratures and the maximum in conjunctions and oppositions is 6° 36' — 4° 56' = 1° 40'. According to Ptolemy, the difference was 2° 40' (HI, 362.1-6). In *De rev.* it is 2° 44' (Th 262.23-32, 265.10-11). Hence I suggest that the figure in our text should be changed from 6° 36' to 7° 36'. Again the reading of S cannot be checked on account of the lacuna in V.

From the following table it can be seen how closely Copernicus adhered to the Ptolemaic determination of the lunar inequalities. The second column contains the maximum inequality in conjunctions and oppositions; the third column shows the greatest additional inequality in the quadratures; and the fourth column sums the second and third.

Ptolemy	5° 0'	2° 40'	7° 40'
Commentariolus	4° 56'	2° 40'	7° 36'
De revolutionibus	4° 56'	2° 44'	7° 40'

Although Ptolemy's table for the first lunar inequality gives 5° 1' as the maximum (HI, 337.21, 390.24), he generally uses the round number 5° in his calculations (HI, 338.22–339.3, 362.1-6, 363.10-12, 364.20-22; cf. Th 257.26-29).

[37] Ptolemy is credited with having discovered the second inequality (HI, 294. 9-14, 354.18–355.20); to account for it, he represented the center of the lunar epicycle as revolving on a circle eccentric to the earth (HI, 355.20-22; cf. Th 232.1-3).

[38] This charge that the representation employed by Ptolemy and his successors violates the axiom of uniform motion is amplified in *De rev.*: "For when they assert that the motion of the center of the epicycle is uniform with respect to the center of the earth, they must also admit that the motion is not uniform on the circle which it describes, namely, the eccentric" (Th 233.11-13). Müller was apparently puzzled by the words *praeter ineptam in ipso circulo motus inaequalitatem* and omitted them from his translation (ZE, XII, 373).

and, in addition, fall into two manifest errors. For the consequence by mathematical analysis is that when the moon is in quadrature, and at the same time in the lowest part of the epicycle, it should appear nearly four times greater (if the entire disk were luminous) than when new and full, unless its magnitude increases and diminishes in no reasonable way.[39] So too, because the size of the earth is sensible in comparison with its distance from the moon, the lunar parallax[40] should increase very greatly at the quadratures. But if anyone investigates these matters carefully, he will find that in both respects the quadratures differ very little from new and full moon, and accordingly will readily admit that my explanation is the truer.

With these three motions in longitude, then, the moon passes through the points of its motion in latitude.[41] The axes of the epicycles are parallel to the axis of the deferent, and therefore the moon does not move out of the plane of the deferent. But the axis of the deferent is inclined to the axis of the great circle

[39] Müller translated the last clause: "es sei denn, man behauptete thörichterweise ein wirkliches Wachsen und Abnehmen der Mondkugel" (unless they absurdly maintained that there is a real increase and decrease in the size of the moon; ZE, XII, 373). This version misses the point. The apparent size of the moon (as measured by its apparent diameter) varies, because the distance of the moon from the earth is not constant (see n. 33). The first of the "two manifest errors" produced by the eccentric is, not that it causes the apparent size of the moon to vary, but that it grossly exaggerates the variation (cf. Th 234.31–235.8).

[40] Müller failed to recognize *diversitas aspectus* as the technical term for parallax (see p. 51, above). Hence he was unable to distinguish the second of the "two manifest errors," and his translation (ZE, XII 373) speaks only of the apparent variation in the size of the moon, "der scheinbare Unterschied in der Grösse." Consequently, in the next sentence, where Copernicus refers to both (*utrumque*) disagreements with the observational data which are produced by the eccentric (1. exaggeration of the variation in the apparent size of the moon; 2. exaggeration of the variation in the lunar parallax), Müller does not know how to render *utrumque*, and falls back on "Grössenunterschied" (variation in size). The explicit statement of Rheticus puts the matter beyond all question: "But experience has shown my teacher that the parallax and size of the moon, at any distance from the sun, differ little or not at all from those which occur at conjunction and opposition, so that clearly the traditional eccentric cannot be assigned to the moon" (p. 134, below).

[41] Here the lacuna in V ends.

or ecliptic;[42] hence the moon moves out of the plane of the ecliptic. Its inclination is determined by the size of an angle which subtends 5° of the circumference of a circle.[43] The poles of the deferent revolve at an equal distance from the axis of the ecliptic,[44] in nearly the same manner as was explained regarding declination.[45] But in the present case they move in the reverse order of the signs and much more slowly, the period of the revolution being nineteen years.[46] It is the com-

[42] Müller rendered *axi magni orbis sive eclipticae* by: "die Achse des grössten Kreises der Ekliptik" (the axis of the great circle of the ecliptic; ZE, XII, 373). This faulty translation shows that Müller did not quite grasp the meaning of *orbis magnus*, which he interpreted (ZE, XII, 365, n. 25) as meaning "great circle" in the geometrical sense, i.e., a circle drawn on the surface of a sphere with its center in the center of the sphere. However, Copernicus's term for "great circle" in the geometrical sense (see above, p. 12, n. 26) is not *orbis magnus* but *circulus maximus* (Th 57-66, *passim*). In the present passage *orbis magnus* bears its usual sense of the real annual revolution of the earth about the sun (see p. 16, above). The *orbis magnus* and the ecliptic lie in the same plane and have a common axis: "But the axis of the great circle is invariably directed toward the points of the firmament which are called the poles of the ecliptic" (p. 64, above).

[43] This estimate of 5° for the maximum latitude of the moon was derived from Ptolemy (HI, 388.11-389.7, 391.52; cf. Th 272.13-15, 274.8-9) and, subject to the correction mentioned in the following note, is retained in modern astronomy.

[44] Therefore the inclination of the moon's orbit to the ecliptic would be constant. That this inclination in fact varies was discovered by Tycho Brahe; see *Tychonis Brahe opera omnia*, ed. Dreyer, II, 121-30, 413.13-21; IV, 42.27-43.22; VI, 170.1-171.8; VII, 151.28-154.35; XI, 162-63; XII, 399-400; Dreyer's remarks on p. liv of the Introduction to Vol. I; and his *Tycho Brahe* (Edinburgh, 1890), pp. 342-44.

[45] See above, p. 64, n. 16. Müller missed the force of *propemodum sicut*, which he translated (ZE, XII, 373) by "ähnlich" (like); whereas "almost like," or something of the sort, is required. In the case of the moon, the poles of the deferent revolve at an equal distance from the axis of the ecliptic, *in aequidistantia axis eclipticae*; but in the case of the motion in declination, the poles of the earth revolve on circles having centers equidistant from the axis of the ecliptic.

[46] This estimate of nineteen years for the period during which the lunar nodes perform their regression was also derived from Ptolemy. He measured the rate of regression by subtracting the moon's mean motion in longitude from the mean motion in latitude (HI, 301.18-23, 356.4-9), the difference being about 3' a day (HI, 356.25-357.6, 358.6-11). By reference to his tables for the moon's motion (HI, 282-293) we can determine the period required for the completion of the circuit as 18 years, 7 months, and 16 days. The discovery that the regression of the nodes is not uniform was made by Tycho Brahe (see the references cited in n. 44, above).

mon opinion that the motion takes place in a higher sphere, to which the poles are attached as they revolve in the manner described. Such a fabric of motions, then, does the moon seem to have.

The Three Superior Planets
Saturn—Jupiter—Mars

Saturn, Jupiter, and Mars have a similar system of motions, since their deferents completely enclose the great circle and revolve in the order of the signs about its center as their common center. Saturn's deferent revolves in 30 years, Jupiter's in 12 years, and that of Mars in 29 months;[47] it is as though the size of the circles delayed the revolutions. For if the radius of the great circle is divided into 25 units, the radius of Mars' deferent will be 38[48] units, Jupiter's 130$\frac{5}{12}$, and Saturn's 230$\frac{1}{6}$.[49] By "radius of the deferent" I mean the distance from the center of the deferent to the center of the first epicycle. Each deferent has two epicycles,[50] one of which carries the

[47] See above, p. 60, n. 7.

[48] Although both S and V read 30, I propose to substitute 38, for the reasons stated in n. 50, below.

[49] S: *230 et sextantem unius*; V: *236 et sextantem unius*. Prowe accepted V, but S is to be preferred, for the reasons given in n. 50, below.

[50] In the *Commentariolus* Copernicus employs for the planets what we have called the concentrobiepicyclic arrangement (see pp. 7, 37, above), consisting of two epicycles upon a deferent which is concentric with the great circle. In *De rev.* this device is replaced, for the three superior planets, by an eccentrepicyclic arrangement, i.e., by a single epicycle upon an eccentric deferent (Th 325.16-21); after indicating the geometric equivalence of the two devices (Th 325.11-16, 327.6-13), Copernicus points to the variation in the eccentricity of the great circle as the reason for his choice of the eccentrepicyclic arrangement (Th 327.13-16). When he wrote the *Commentariolus*, he regarded this eccentricity as constant (see above, p. 61, n. 9).

Now if the two arrangements are to produce identical results, then, as Copernicus points out, the radius (R) of the concentric deferent (*Commentariolus*) must be equal to the radius (R) of the eccentric deferent (*De rev.*). Let r denote the radius of the great circle. By a comparison of the ratio R:r, as given here, with the values in *De rev.*, we may discover whether in shifting from the concentrobiepicyclic to the eccentrepicyclic arrangement Copernicus altered the relative sizes of the deferent and great circle. In *De rev.*, for Saturn r = 1090 (Th 341.29), for Jupiter r = 1916 (Th 353.15-16), and for Mars r = 6580 (Th 364.8-9), when in each case R = 10,000.

other, in much the same way as was explained in the case of the moon,[51] but with a different arrangement. For the first epicycle revolves in the direction opposite to that of the deferent, the periods of both being equal. The second epicycle, carrying the planet, revolves in the direction opposite to that of the first with twice the velocity. The result is that whenever the second epicycle is at its greatest or least distance from the center of the deferent, the planet is nearest to the center of the first epicycle; and when the second epicycle is at the midpoints, a quadrant's distance from the two points just men-

---

	R:r *De revolutionibus*	*Commentariolus*
Saturn	10,000:1090 = 229⅓:25	230⅙:25
Jupiter	10,000:1916 = 130½:25	130 5/12:25
Mars	10,000:6580 = 38:25	38:25

The table enables us to deal with a variant reading in this passage. For R in the case of Saturn, S has 230⅙, while V gives 236⅙ (Curtze's collation [MCV, IV, 7] inaccurately assigns to Jupiter the reading of S for Saturn). Prowe accepted the reading of V, but S is clearly preferable, as the following analysis will show.

I have already referred (see p. 69, n. 30) to the series of notes made by Copernicus in his copy of the Tables of Regiomontanus. Curtze correctly pointed out that the ratios contained in these notes are identical with those adopted in the *Commentariolus* (MCV, IV, 7 n); and he used the statement about the moon to emend a false reading in our text. However, he failed to make any further use of these entries. Now for the radius (not diameter, as MCV, I, 12 n. and PII, 195 n. have it; cf. PII, 211) of Saturn's deferent, they give 230⅚ (not 230⅝, as PII, 195 n). Hence we are justified in preferring the reading of S to that of V. This judgment is confirmed by the fact that Tycho Brahe's reference to the *Commentariolus* agrees with S (see his *Opera omnia*, ed. Dreyer, II, 428.40–429.2).

Moreover, these notes of Copernicus show that S and V agree on a false reading for R in the case of Mars. The statement in the Tables of Regiomontanus gives the radius of Mars' deferent as approximately 38 (*Martis semidiametrus orbis 38 fere*). Now a value of 30, which is the reading of both our MSS, would make the ratio R:r for Mars 30:25 = 10,000:8333, at wide variance from the corresponding ratio in *De rev.* But reference to the table will show that the agreement between *De rev.* and the *Commentariolus* for both Saturn and Jupiter is quite close. Hence I have adopted 38, the number written by Copernicus in his Tables of Regiomontanus, in place of 30. Writing o for 8 is not an uncommon error of copyists (cf. below, p. 82, n. 74).

[51] See the opening paragraph of the section on "The Moon."

tioned,[52] the planet is most remote from the center of the first epicycle. Through the combination of these motions of the deferent and epicycles, and by reason of the equality of their revolutions, the aforesaid withdrawals and approaches occupy absolutely fixed places in the firmament, and everywhere exhibit unchanging patterns of motion. Consequently the apsides are invariable;[53] for Saturn, near the star which is said to be on the elbow of Sagittarius;[54] for Jupiter, 8° east of the star which is called the end of the tail of Leo;[55] and for Mars, 6½° west of the heart of Leo.[56]

[52] Müller rendered *in quadrantibus autem mediantibus* by: "zur Zeit der mittleren Quadraturen" (at the time of the mean quadratures; ZE, XII, 374). With regard to *mediantibus*, this version repeats the blunder pointed out above in n. 29 on p. 69; and, in addition, it mistakes *quadrans* (quadrant, the fourth part of a circumference) for *quadratura* (quadrature; cf. above, p. 47, n. 163).

[53] This was Ptolemy's view. He held that the planetary apsides were fixed in relation to the sphere of the fixed stars, since, as measured by the equinoxes and solstices, both the apsides and the fixed stars moved in the same direction at the same slow rate (HII, 251.24–252.7, 252.11-18, 257.3-12, 269.3-11; cf. Th 308.20-24).

[54] The star is here described as *quae super cubitum esse dicitur Sagittatoris*. It is unquestionably to be identified with Sagittarius 19 in Ptolemy's catalogue (HII, 114.10), for that star was described in the first printed translation of the *Syntaxis* into Latin (Venice, 1515, p. 84r) as *quae est super cubitum dextrum*. In *De rev.* Copernicus uses instead the name *In dextro cubito* (Th 139.14).

[55] This star is Leo 27 in Ptolemy's catalogue and in *De rev.* (HII, 100.7; Th 135.12). Its Bayer name is β Leonis.

[56] This star is Leo 8 (HII, 98.6; Th 134.23-24). It was called Basiliscus or Regulus, and its Bayer name is α Leonis.

Ptolemy had put the apogee of Saturn at 23° of Scorpio; of Jupiter, at 11° of Virgo; and of Mars, at 25° 30′ of Cancer (HII, 412.12-17, 380.22-381.4, 345.12-20). In his catalogue of the fixed stars these places are, respectively, 31° 50′ west of Sagittarius 19, 16° 30′ east of Leo 27, and 7° west of Leo 8. From them Copernicus's determinations differ, respectively, by 31° 50′ eastward, 8° 30′ westward, and ½° eastward. Hence we may say that although in the *Commentariolus* Copernicus accepted Ptolemy's doctrine of the fixity of the planetary apsides, he intended to put forward improved determinations of them.

In *De rev.* the places are again altered. But now they are all east of Ptolemy's determinations; for Saturn's apogee is 17° 49′ west of Sagittarius 19; Jupiter's, 21° 10′ east of Leo 27; and Mars', 3° 50′ east of Leo 8 (Th 338.15-18, 350.15-16, 360.3-5). Hence Copernicus abandons the idea of the fixed apogee and enunciates the discovery that the longitude of the planetary apogees increases: "Moreover, the position of the higher apse of [Saturn's] eccentric has in the meantime advanced 13° 58′ in the sphere of the fixed stars. Ptolemy believed that this posi-

The radius of the great circle was divided above into 25 units. Measured by these units, the sizes of the epicycles are as follows. In Saturn the radius of the first epicycle consists of 19 units, 41 minutes; the radius of the second epicycle, 6 units, 34 minutes. In Jupiter the first epicycle has a radius of 10 units, 6 minutes; the second, 3 units, 22 minutes. In Mars the first epicycle, 5 units, 34 minutes; the second, 1 unit, 51 minutes.[57] Thus the radius of the first epicycle in each case is three times as great as that of the second.[58]

The inequality which the motion of the epicycles imposes upon the motion of the deferent is called the first inequality; it follows, as I have said, unchanging paths everywhere in the firmament. There is a second inequality, on account of which the planet seems from time to time to retrograde, and often to become stationary. This happens by reason of the motion, not of the planet, but of the earth changing its position in the great circle. For since the earth moves more rapidly than the planet, the line of sight directed toward the firmament regresses, and the earth more than neutralizes the motion of the planet. This regression is most notable when the earth is nearest to the planet, that is, when it comes between the sun and the planet at the evening rising of the planet. On the other hand, when

----

[57] I resume the comparison instituted above in n. 50 on p. 74. As Copernicus points out (Th 327.7-8), the radius (E) of the first epicycle (*Commentariolus*) must be equal to the eccentricity (E) of the eccentric (*De rev.*). Now in *De rev.* for Saturn E = 854 (Th 330.18), for Jupiter E = 687 (Th 343.23-28), and for Mars E = 1460 (Th 358.28); we already have the values of r.

r:E

	De revolutionibus	*Commentariolus*
Saturn	1090: 854 = 25:19^{p}35^m	25:19^{p}41^m
Jupiter	1916: 687 = 25: 8^{p}58^m	25:10^p 6^m
Mars	6580:1460 = 25: 5^{p}33^m	25: 5^{p}34^m

[58] Hence the radius of the second epicycle in the *Commentariolus* is equal to the radius of the single epicycle in *De rev.*, since both = ⅓ E (Th 325.19-20). An exception will be noted in the case of Mars, where Copernicus reduces the eccentricity from 1,500 (Th 354.29-355.2) to 1,460, but leaves the radius of the epicycle at 500 (Th 358.24-31, 360.7-11, 362.26-28).

the planet is setting in the evening or rising in the morning, the earth makes the observed motion greater than the actual. But when the line of sight is moving in the direction opposite to that of the planets and at an equal rate, the planets appear to be stationary, since the opposed motions neutralize each other; this commonly occurs when the angle at the earth between the sun and the planet is 120°.[59] In all these cases, the lower the deferent on which the planet moves, the greater is the inequality. Hence it is smaller in Saturn than in Jupiter, and again greatest in Mars, in accordance with the ratio of the radius of the great circle to the radii of the deferents. The inequality attains its maximum for each planet when the line of sight to the planet is tangent to the circumference of the great circle. In this manner do these three planets move.

In latitude they have a twofold deviation. While the circumferences of the epicycles remain in a single plane with their deferent, they are inclined to the ecliptic. This inclination is governed by the inclination of their axes, which do not revolve, as in the case of the moon,[60] but are directed always toward the same region of the heavens. Therefore the intersections of the deferent and ecliptic (these points of intersection are called the nodes) occupy eternal places in the firmament.[61] Thus the node where the planet begins its ascent toward the north is, for Saturn, 8½° east of the star which is described as being in the head of the eastern of the two Gemini;[62] for Jupiter, 4° west of

[59] Cf. Pliny *Natural History* ii.15(12).59: "In the trine aspect, that is, at 120° from the sun, the three superior planets have their morning stations, which are called the first stations . . . and again at 120°, approaching from the other direction, they have their evening stations, which are called the second stations"; cf. also ii.16(13).69-71. It has been shown that Copernicus read carefully a copy of the Rome, 1473 edition of Pliny's *Natural History* (L. A. Birkenmajer, *Stromata Copernicana*, Cracow, 1924, pp. 327-34); and also a copy of the Venice, 1487 edition (MCV, I, 40-41).

[60] See the closing paragraph of the section on "The Moon."

[61] Copernicus derived from Ptolemy the view that the nodes, like the apsides, are fixed (HII, 530.8-11; cf. Karl Manitius, *Des Claudius Ptolemäus Handbuch der Astronomie*, Leipzig, 1912-13, II, 426). But in *De rev.*, having discovered the motion of the apsides, Copernicus holds that this motion is shared by the nodes (Th 413.7-15, 415.20-25).

[62] Gemini 2 (HII, 92.4; Th 132.31-32), Pollux, β Geminorum; cf. above, p. 62, n. 12.

114

the same star; and for Mars, $6\frac{1}{2}°$ west of Vergiliae.[63] When the planet is at this point and its diametric opposite, it has no latitude. But the greatest latitude, which occurs at a quadrant's distance from the nodes,[64] is subject to a large inequality. For the inclined axes and circles seem to rest upon the nodes, as though swinging from them. The inclination becomes greatest when the earth is nearest to the planet, that is, at the evening rising of the planet; at that time the inclination of the axis is, for Saturn $2\frac{2}{3}°$, Jupiter $1\frac{2}{3}°$, and Mars $1\frac{5}{6}°$.[65] On the other hand, near the time of the evening setting and morning rising, when the earth is at its greatest distance from the planet, the inclination is smaller,[66] for Saturn and Jupiter by $\frac{5}{12}°$, and for

[63] Taurus 30 (HII, 90.2; Th 132.5-6). Authorities differ about the identification of Taurus 30; see Christian H. F. Peters and Edward B. Knobel, *Ptolemy's Catalogue of Stars* (Carnegie Institution of Washington, Publication No. 86, 1915), p. 115.

[64] Ptolemy had put the points of greatest northern latitude for Saturn and Jupiter at 0° of Libra, and for Mars at 30° of Cancer (HII, 526.6-11; cf. Th 413.7-11). If we compare these places with his determinations of the apogees (see above, p. 76, n. 56), we find that for Saturn the point of greatest northern latitude is 53° west of the apogee; for Jupiter, 19° east; and for Mars, 4° 30' east. Ptolemy states these differences of position in round numbers as 50° west, 20° east, and 0° (HII, 587.5-9; cf. Manitius, *Ptolemäus Handbuch*, II, 425, n. 21).

In the present passage Copernicus gives the places of the ascending nodes. By adding 90° to these places, we obtain the points of greatest northern latitude. They turn out to be, for Saturn, 79° 40' west of the apogee; for Jupiter, 20° 10' east; and for Mars, 0° 20' west. In the *Commentariolus*, then, Copernicus not only adheres to the Ptolemaic ideas of the fixed apogee and the fixed node, but he also retains Ptolemy's distance between apogee and node for Jupiter and Mars, although increasing the distance by 30° for Saturn.

In *De rev.*, although the apogee moves, the distance between apogee and node remains constant, since the node shares the motion of the apogee. Copernicus finds the points of greatest northern latitude, for Saturn at 7° of Scorpio; for Jupiter at 27° of Libra; and for Mars at 27° of Leo (Th 413.11-13). If we compare these places with his determinations of the apogees (see above, p. 76, n. 56), we find that for Saturn the point of greatest northern latitude is 23^b 21' west of the apogee; for Jupiter, 48° east; and for Mars, 27° 20' east.

[65] S: *dextante*; V: *sextante*. Prowe, followed by Müller (ZE, XII, 377), adopted the reading of V; but *sextante* is clearly impossible, for the following sentence of the text states that the inclination diminishes in the case of Mars by $1\frac{2}{3}°$.

[66] The inclination is greatest when the planet is in opposition, smallest when the planet is in conjunction; and the greatest difference between maximum and

115

Mars by $1\frac{2}{3}°$. Thus this inequality is most notable in the greatest latitudes, and it becomes smaller as the planet approaches the node, so that it increases and decreases equally with the latitude.

The motion of the earth in the great circle also causes the observed latitudes to change, its nearness or distance increasing or diminishing the angle of the observed latitude, as mathematical analysis demands. This motion in libration occurs along a straight line, but a motion of this sort can be derived from two circles. These are concentric, and one of them, as it revolves, carries with it the inclined poles of the other. The lower circle revolves in the direction opposite to that of the upper, and with twice the velocity. As it revolves, it carries with it the poles of the circle which serves as deferent to the epicycles. The poles of the deferent are inclined to the poles of the circle

---

minimum occurs at the points of greatest latitude (Th 415.9-14). The following table compares the maximum and minimum angles of inclination as given here with those in *De rev.* (Th 421.22-25; 421.31–422.1; 422.7-8, 10-11).

	Angles of Inclination	Commentariolus	De revolutionibus
Saturn	Greatest	2° 40′	2° 44′
	Least	2° 15′	2° 16′
Jupiter	Greatest	1° 40′	1° 42′
	Least	1° 15′	1° 18′
Mars	Greatest	1° 50′	1° 51′
	Least	0° 10′	0° 9′

From the table we see that the main inclinations and their limits of variation are as follows:

	Commentariolus	De revolutionibus
Saturn	2° 27½′ ± 12½′	2° 30′ ± 14′
Jupiter	1° 27½′ ± 12½′	1° 30′ ± 12′
Mars	1° ± 50′	1° ± 51′

In Ptolemy's treatment of the latitudes, for the three superior planets the angle at which the eccentric deferent was inclined to the ecliptic was constant (HII, 529.3-9). His values were: for Saturn 2° 30′, for Jupiter 1° 30′, and for Mars 1° (HII, 540.13-14, 542.5-9). But the epicycle was inclined to the eccentric at a varying angle (HII, 529.12–530.8). It will be observed that in Copernicus's theory the epicycles and deferent are coplanar; hence the angle at which the deferent is inclined to the ecliptic cannot be fixed, but must vary (Th 413.1-3, 29-31).

116

halfway[67] above at an angle equal to the inclination of these poles to the poles of the highest circle.[68] So much for Saturn, Jupiter, and Mars and the spheres which enclose the earth.

Venus

There remain for consideration the motions which are included within the great circle, that is, the motions of Venus and Mercury. Venus has a system of circles like the system of the superior planets,[69] but the arrangement of the motions is different. The deferent revolves in nine months, as was said above,[70] and the greater epicycle also revolves in nine months. By their composite motion the smaller epicycle is everywhere brought back to the same path in the firmament, and the higher apse is at the point where I said the sun reverses its course.[71] The period of the smaller epicycle is not equal to that of the deferent and greater epicycle,[72] but has a constant relation to

[67] S: *mediate*; V: *mediale*. Before S was known, Curtze emended V to *immediate*, which Prowe prints. But S is undoubtedly correct.

[68] Since motion in a straight line would violate the principle of circularity, Copernicus is at pains to prove that a rectilinear motion may be produced by a combination of two circular ones. A less concise account of this geometric device, employed in connection with the theory of precession, as well as an explanation of the term "libration," will be found in the *Narratio prima* (pp. 153-54, below; cf. Th 165.18–169.22).

[69] In *De rev.* Copernicus replaces the concentrobiepicyclic arrangement for Venus by an eccentreccentric arrangement, i.e., by two eccentrics (Th 368.23-29). The larger, outer eccentric which carries the planet has for its center a point which revolves on the smaller eccentric (Th 368.30–369.6).

[70] Page 60.

[71] In placing the apogee of Venus at the solar apogee Copernicus retains the Ptolemaic idea of the fixed apse, but he offers an improved determination. For Ptolemy had put the apogee of Venus at 25° of Taurus (HII, 300.15-16; cf. Th 365.20-25; 366.3-7, 17-20), and the solar apogee at 5° 30′ of Gemini (see above, p. 62, n. 13). Hence for him the apogee of Venus was 10° 30′ west of the solar apogee. Now we have already seen that in the *Commentariolus* Copernicus advances the solar apogee 11° 10′, as measured by the fixed stars, over Ptolemy's determination. Hence he advances the apogee of Venus 21° 40′, again as measured by the fixed stars, over Ptolemy's determination.

[72] This is the difference between the arrangement of the motions, on the one hand, for the three superior planets, and on the other hand, for Venus. In the former case the period of the smaller epicycle is one-half the period of the deferent and greater epicycle (see the opening paragraph of the section on "The

the motion of the great circle. For one revolution of the latter the smaller epicycle completes two. The result is that whenever the earth is in the diameter drawn through the apse, the planet is nearest to the center of the greater epicycle; and it is most remote, when the earth, being in the diameter perpendicular to the diameter through the apse, is at a quadrant's distance from the positions just mentioned. The smaller epicycle of the moon moves in very much the same way with relation to the sun.[73] The ratio of the radius of the great circle to the radius of the deferent of Venus is 25:18;[74] the greater epicycle has a value of ¾ of a unit, and the smaller ¼.[75]

Three Superior Planets"). Müller completely missed the distinction. His translation runs: "Die Umlaufszeit dieses kleineren Epicykels ist verschieden von der der oben genannten Kreise; so entsteht längst der Ekliptik eine ungleichförmige Bewegung. Vollführen jene einen Umlauf, so führt der kleinere einen doppelten aus" (The period of this smaller epicycle is different from that of the above-mentioned circles [i.e., deferent and greater epicycle]; thus there appears along the ecliptic an unequal motion. While those circles [i.e., deferent and greater epicycle] complete one revolution, the smaller epicycle completes two; ZE, XII, 378). The source of Müller's difficulty seems to have been the unusual expression *Minor autem epicyclus impares cum illis revolutiones habens, motui orbis magni imparitatem reservavit.* This may be literally translated as follows: "The smaller epicycle, having revolutions unequal with those of the deferent and greater epicycle, has reserved the inequality for the motion of the great circle." The next sentence in the text makes Copernicus's meaning clear beyond dispute. The revolution of the smaller epicycle takes half the time required by the motion on the great circle.

[73] See the opening paragraph of the section on "The Moon."

[74] S has the false reading 10, instead of 18 (Lindhagen reproduces this page of the MS). I call attention to the copyist's error of writing o for 8, in connection with the emendation proposed in the last paragraph of n. 50 (p. 75, above).

[75] To discover whether in shifting from the concentrobiepicyclic arrangement in the *Commentariolus* to the eccentreccentric arrangement in *De rev.* Copernicus altered the relative sizes of the circles, we may make the following comparisons. The radius (R) of the concentric deferent (*Commentariolus*) corresponds to the radius (R) of the outer eccentric (*De rev.*). Similarly, the radius (E) of the first epicycle (*Commentariolus*) corresponds to the eccentricity (E) of the outer eccentric (*De rev.*); and since the eccentricity varies, we take its mean value. Let r denote the radius of the great circle. Now in *De rev.* R = 7193, r = 10,000, and E = 312 (Th 367.13-14, 368.12-22, 371.11). Then in *De rev.* r:R = 10,000:7193 = 25:17.98, while in the *Commentariolus* r:R = 25:18; in *De rev.* r:E = 10,000:312 = 25:0.78, while in the *Commentariolus* r:E =

Venus seems at times to retrograde, particularly when it is nearest to the earth, like the superior planets, but for the opposite reason. For the regression of the superior planets happens because the motion of the earth is more rapid than theirs, but with Venus, because it is slower; and because the superior planets enclose the great circle, whereas Venus is enclosed within it. Hence Venus is never in opposition to the sun, since the earth cannot come between them, but it moves within fixed distances on either side of the sun. These distances are determined by tangents to the circumference drawn from the center of the earth, and never exceed 48° in our observations.[76] Here ends the treatment of Venus' motion in longitude.

Its latitude also changes for a twofold reason. For the axis of the deferent is inclined at an angle of 2½°,[77] the node whence the planet turns north being in the apse. However, the deviation which arises from this inclination, although in itself it is one and the same, appears twofold to us.[78] For when the earth is on the line drawn through the nodes of Venus, the deviations on the one side are seen above, and on the opposite

---

25:0.75. The radius of the second epicycle = ⅓ E, a ratio which is applied in the *Commentariolus* to all the planets. In *De rev.* the radius of the smaller eccentric, being one-third of the mean eccentricity of the outer eccentric (Th 368.18-22), also = ⅓ E. Hence the second epicycle (*Commentariolus*) corresponds to the smaller eccentric (*De rev.*).

Despite *dodrantem* in the text, Müller's translation makes E = ⅔ (ZE, XII, 378). He was evidently confused by a misprint in Prowe's footnote (PII, 198). Yet in that same footnote, five lines below the misprint, the correct value of ¾ appears (cf. PII, 211 and MCV, I, 14-15 n).

[76] This value of 48° for the greatest elongation of Venus was derived from Ptolemy (HII, 522.14), and is accepted by modern astronomy.

[77] Müller wrote 2° (ZE, XII, 379). He was evidently unfamiliar with *s.* as the abbreviation of *semissis*, "one-half" (cf. Th 71.23, 167.4, 425.25). For in his note on the matter he misinterpreted *s.* as the abbreviation of *scrupula*, "minutes" (this word was not assigned to the masculine gender, as Müller thought). Had he consulted Curtze's collation of S and V, his difficulty would have been obviated. For Curtze, confronted by a variant reading (MCV, IV, 8), showed that 2½° is supported by *De rev.* (Th 424.23-24). Moreover, in Ptolemy's treatment of the latitude of Venus, there are two inclinations of the epicycle, and each is given as 2½° (HII, 535.15-18, 536.8-11).

[78] S, V: *duplex non ostenditur*. Müller correctly emended to *duplex nobis ostenditur* (ZE, XII, 379, n. 72).

side below; these are called the reflexions.[79] When the earth is at a quadrant's distance[80] from the nodes, the same natural inclinations of the deferent appear, but they are called the declinations. In all the other positions of the earth, both latitudes mingle and are combined, each in turn exceeding the other; by their likeness and difference they are mutually increased and eliminated.

The inclination of the axis is affected by a motion in libration that swings, not on the nodes as in the case of the superior planets,[81] but on certain other movable points. These points perform annual revolutions with reference to the planet. Whenever the earth is opposite the apse of Venus, at that time the amount of the libration attains its maximum for this planet, no matter where the planet may then be on the deferent. As a consequence, if the planet is then in the apse or diametrically opposite to it, it will not completely lack latitude, even though it is then in the nodes. From this point the amount of the libration decreases, until the earth has moved through a quadrant of a circle from the aforesaid position, and, by reason of the likeness of their motions, the point of maximum deviation[82] has moved an equal distance from the planet. Here no trace of the deviation is found.[83] Thereafter the descent of the deviation continues.[84] The initial point drops from north to south,

---

[79] An alternative name was obliquation: "They call this deviation of the planet the obliquation, but some call it the reflexion" (Th 418.22-23). In *De rev.* Copernicus generally uses obliquation, but in the *Narratio prima* Rheticus favors reflexion.

[80] Müller's translation: "in den Quadraturen" (in the quadratures; ZE, XII, 379) again confuses *quadrantibus* with *quadraturis* (cf. above, p. 76, n. 52). An inferior planet cannot come to quadrature (see above, p. 50); Copernicus has just stated that the greatest elongation of Venus is 48°.

[81] See the penultimate paragraph of the section on "The Three Superior Planets."

[82] Müller correctly emended *maxime* (S, V) to *maximae* (ZE, XII, 380, n. 75).

[83] Since the deviation vanishes when the earth is 90° from the apse-line of the planet, the deviation has no effect upon the declinations, but only upon the reflexions. Copernicus employs the deviation "because the angle of inclination . . . is found to be greater in the obliquation [reflexion] than in the declination" (Th 418.27-29).

[84] S, V: *continuato*. Prowe's *continuatio* is a misprint (PII, 200.3).

constantly increasing its distance from the planet in accordance with the distance of the earth from the apse. Thereby the planet is brought to the part of the circumference which previously was south. Now, however, by the law of opposition, it becomes north and remains so until the limit of the libration is again reached upon the completion of the circle. Here the deviation becomes equal to the initial deviation and once more attains its maximum. Thus the second semicircle is traversed in the same way as the first. Consequently this latitude, which is usually called the deviation, never becomes a south latitude. In the present instance, also, it seems reasonable that these phenomena should be produced by two concentric circles with oblique axes, as I explained in the case of the superior planets.[85]

Mercury

Of all the orbits in the heavens the most remarkable is that of Mercury, which traverses almost untraceable paths, so that it cannot be easily studied. A further difficulty is the fact that the planet, following a course generally invisible in the rays of the sun, can be observed for a very few days only. Yet Mercury too will be understood, if the problem is attacked with more than ordinary ability.

Mercury, like Venus, has two epicycles which revolve on the deferent.[86] The periods of the greater epicycle and deferent are equal, as in the case of Venus. The apse is located $14\frac{1}{2}°$ east of Spica Virginis.[87] The smaller epicycle revolves with twice the velocity of the earth. But by contrast with Venus, whenever the earth is above the apse or diametrically opposite

[85] For a fuller account of Copernicus's theory for the latitudes of Venus see pp. 180-85, below.

[86] In De rev. Copernicus replaces the concentrobiepicyclic arrangement for Mercury by an eccentreccentric arrangement (Th 377.2-3).

[87] Since Ptolemy had put the apogee of Mercury at 10° of Libra (HII, 264.12-14, 271.2-4; cf. Th 380.6-7), and Spica at 26° 40' of Virgo (HII, 103.16), the apse was 13° 20' east of Spica Virginis. Hence in the Commentariolus Copernicus retains the idea of the fixed apse and modifies its position slightly. But in De rev. he puts the apse 41° 30' east of Spica (Th 136.10, 389.5-6, 393.5-8), and extends to Mercury the principle that the longitude of the planetary apogees increases (Th 393.16-19, 27-29; cf. n. 56 on pp. 76-77, above).

to it, the planet is most remote from the center of the greater epicycle; and it is nearest, whenever the earth is at a quadrant's distance[88] from the points just mentioned. I have said[89] that the deferent of Mercury revolves in three months, that is, in 88 days. Of the 25 units into which I have divided the radius of the great circle, the radius of the deferent of Mercury contains 9⅖. The first epicycle contains 1 unit, 41 minutes; the second epicycle is ⅓ as great, that is, about 34 minutes.[90]

But in the present case this combination of circles is not sufficient, though it is for the other planets. For when the earth passes through the above-mentioned positions with respect to the apse the planet appears to move in a much smaller path[91] than is required by the system of circles described above; and in a much greater path,[91] when the earth is at a quadrant's distance[92] from the positions just mentioned. Since no other inequality in longitude is observed to result from this, it may be reasonably explained by a certain approach of the planet to and withdrawal from the center of the deferent[93] along a

[88] Again Müller erroneously translates by "in the quadratures" (ZE, XII, 381). Mercury, like Venus, cannot come to quadrature (cf. above, p. 84, n. 80).

[89] Page 60, above.

[90] The analysis made above (p. 82, n. 75) for Venus is equally applicable here. In De rev. R (mean value) = 3,763, r = 10,000, and E (mean value) = 736 (Th 382.9-10, 382.27-383.2). Then in De rev. r:R = 10,000:3763 = 25: 9.41, while in the Commentariolus r:R = 25:9.40; in De rev. r:E = 10,000:736 = 25:1.84, while in the Commentariolus r:E = 25:1.68. The radius of the second epicycle (Commentariolus) = ⅓ E. But the radius of the smaller eccentric (De rev.) = ⅓ E, only where E denotes the eccentricity of the outer eccentric (Th 377.11-15), as set down in conformity with the general planetary theory used in De rev. As in the case of Mars (see above, p. 77, n. 58), Copernicus modifies the ratio; the radius of the smaller eccentric = 212 (Th 382.8-9), or 2/7 E, where E denotes the mean eccentricity of the outer eccentric.

[91] Müller translates longe minori apparet ambitu sidus moveri by: "so scheint der Planet sich viel langsamer zu bewegen" (the planet appears to move much more slowly); and longe etiam maiore by: "viel schneller" (much more swiftly; ZE, XII, 381). However, Copernicus is concerned here with the variations, not in Mercury's velocity, but in its distance from the center of the great circle.

[92] Failing to recognize that quadratura is used here and again near the close of this paragraph in the sense of "quadrant" (see above, p. 47, n. 163), Müller inaccurately translates by "in the quadratures" (ZE, XII, 381, 382).

[93] S: a centro orbis; V: centri orbis. Prowe accepted V, although S is certainly correct.

straight line. This motion must be produced by two small circles stationed about the center of the greater epicycle, their axes being parallel to the axis of the deferent. The center of the greater epicycle, or of the whole epicyclic structure, lies on the circumference of the small circle that is situated between this center and the outer small circle. The distance from this center to the center of the inner circle is exactly[94] equal to the distance from the latter center to the center of the outer circle.[95] This distance has been found to be 14½ minutes[96] of one unit of the

[94] S, V: *asse*. For this sound reading Curtze incorrectly substituted *axe* (MCV, I, 17.5), which Prowe accepted (PII, 201.10). By ignoring the rules of syntax Müller contrived to incorporate *axe* in his translation.

[95] Let the dotted circumference (Fig. 24) represent the inner small circle with

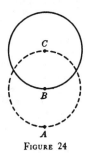

FIGURE 24

its center at B; and the unbroken circumference, the outer small circle with its center at C. The center of the greater epicycle is at A; and AB = BC.

[96] S: *minut. 14 et medio;* V: *minutibus 24 et medio.* Prowe accepted V, although S is certainly correct, as the closing words of this paragraph show. Again Copernicus's notations in his copy of the Tables of Regiomontanus aid us. For the entry concerning Mercury gives values for the deferent and epicycles that agree with those in our text. Then it adds that the inequality of the diameter is 29 minutes (*diversitas diametri 0.29*). Now Curtze, Prowe, and Müller quoted the entry in their notes (MCV, I, 16 n; PII, 201 n; ZE, XII, 381, n. 78). All three called attention to the agreement between the entry and our text with reference to the deferent and epicycles. But they failed to see that the "approach and withdrawal" of our text is identical with the "inequality of the diameter" in the entry; and that the value of 29 minutes given in both places establishes the correctness of S as against V.

This value varies but slightly from Ptolemy's. In his system, the inequality is produced by a small circle upon which the center of the eccentric revolves (HII, 252.26–253.6, 256.15-22; cf. Th 376.17-24). If we compare the radius of the small circle with the sum of the radii of the eccentric and epicycle (HII, 279.15-

25 by which I have measured the relative sizes of all the circles. The motion of the outer small circle performs two revolutions in a tropical year,[97] while the inner one completes four in the same time with twice the velocity in the opposite direction. By this composite motion the centers of the greater epicycle are carried along a straight line, just as I explained with regard to the librations in latitude.[98] Therefore, in the aforementioned positions of the earth with respect to the apse, the center of the greater epicycle is nearest to the center of the deferent; and it is most remote, when the earth is at a quadrant's distance[92] from these positions. When the earth is at the midpoints, that is, 45° from the points just mentioned, the center of the greater epicycle joins the center of the outer[99] small circle, and both centers coincide.[100] The amount of this with-

---

18), we get the ratio 1:27½, while in the *Commentariolus* the corresponding ratio is 1 : 24 (29^m : 9^{p}24^m + 1^{p}41^m + 34^m).

In *De rev.* Copernicus represents the inequality by adding an epicycle to the outer eccentric (Th 377.4-8, 18-23); so that, if we include this refinement, his arrangement for Mercury in *De rev.* is bieccentrepicyclic rather than eccentreccentric (Th 377.23-26). But he does not alter the amount of the inequality. For he puts the diameter of the epicycle at 380, where r = 10,000 (Th 382.23-27, 384.9-14). Then the amount of the inequality is 190 (r = 10,000), or 28½ minutes, where r = 25.

[97] Müller failed to recognize *annus vertens* as the term for "tropical year" (see p. 46, above).

[98] See the closing paragraph of the section on "The Three Superior Planets."

[99] Müller omitted *exterioris* from his translation (ZE, XII, 382).

[100] Figure 25 may serve to clarify this motion in libration. In the initial position, the earth is at E1 on the produced apse-line, the center of the greater epicycle is at A, the center of the inner small circle is at B1, and the center of the outer small circle is at C. While the earth moves 45° from E1 to E2, the outer circle rotates through a quadrant, thereby moving the center of the inner circle from B1 to B2. But during this interval, the inner circle rotates through a semicircle, thereby bringing the center of the epicycle to C. As the earth moves 45° from E2 to E3, the center of the inner circle reaches B3, and the center of the epicycle comes to D. As the earth moves from E3 to E4, the center of the inner circle goes to B4, and the center of the epicycle to C. When the earth arrives at E5, the center of the inner circle returns to B1, and the center of the epicycle to A. While the earth completes the remaining semicircle E5-E6-E7-E8-E1, the small circles repeat their previous motion. Therefore, whenever the earth is on the produced apse-line (E1 or E5), the center of the greater epicycle is nearest (A) to the center of the deferent. When the earth is at a quadrant's distance from the apse-

drawal and approach is 29 minutes[96] of one of the above-mentioned units. This, then, is the motion of Mercury in longitude.

Its motion in latitude is exactly like that of Venus, but always in the opposite hemisphere. For where Venus is in north latitude, Mercury is in south. Its deferent is inclined to the ecliptic at an angle of 7°.[101] The deviation, which is always south, never

line (E3 or E7), the center of the epicycle is most remote (D) from the center of the deferent. When the earth is at E2, E4, E6, or E8, the center of the epicycle coincides with C, the center of the outer small circle.

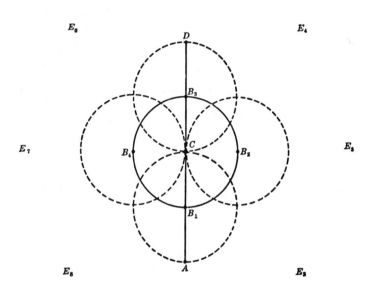

FIGURE 25

[101] In *De rev.* the angle is given as 6° 15' for the declinations, and 7° for the reflexions (Th 424.23-27, 431.4-9). These were Ptolemy's values (HII, 536.20-22, 575.9-11).

exceeds ¾°.[102] For the rest, what was said about the latitudes of Venus may be understood here also, to avoid repetition.

Then Mercury runs on seven circles in all; Venus on five; the earth on three, and round it the moon on four; finally Mars, Jupiter, and Saturn on five each. Altogether, therefore, thirty-four circles suffice to explain the entire structure of the universe and the entire ballet of the planets.

[102] This was the traditional estimate (Th 433.20-25); but in *De rev.* Copernicus puts it at 51′ ± 18′ (Th 435.4-8; 440-41). Müller rendered this sentence by: "doch übersteigt die Ablenkung nach Süden nie den zwölften Teil eines Grades" (the southward deviation never exceeds ½2°; ZE, XII, 382). This version omits *semper*, and puts *dodrantem* = ½2. For ½2 Copernicus wrote the usual word *uncia* (Th 159.28).

4

Galileo Galilei, THE STARRY MESSENGER
LETTER TO THE GRAND DUCHESS

(Trans. by Stillman Drake)

1. Why were the discoveries which Galileo made with the telescope so astounding? What confirmation of Copernicus' theories did they offer?

2. How are Galileo's discoveries a good example of the use of a scientific method similar to that proposed by Bacon? Does Galileo's method surpass Bacon's in any ways?

3. Why did Galileo name the newly discovered moons of Jupiter "the Medicean stars"? How was Galileo's research financed? How do you think theoretical scientific research should be funded? Should public money be made available to support scientific research?

4. Why did Galileo encounter so much opposition when he supported Copernicus' doctrine, even though this doctrine had been favorably received by the church when Copernicus published it half a century earlier?

5. Why was the assertion that the earth moves so disturbing to Galileo's opponents? What religious arguments did they offer against this assertion? Were there more important underlying reasons for their opposition? Do you think any similar opposition to science is alive today?

6. Does Galileo find any contradiction between religion and science? How does he explain the apparent contradiction? Do you think that there is any contradiction?

7. How does Galileo's experience raise the issue of freedom of intellectual inquiry? Is this issue still an important one today?

The son of a Florentine cloth merchant, Galileo Galilei (1564-1642) studied and taught mathematics, physics, and astronomy. His astronomical observations by means of the telescope and his physical experiments concerning motion were of profound significance in the scientific revolution. His writings were aimed at a fairly wide audience. Forced to defend himself against the fierce opposition of some churchmen and especially against Aristotelian philosophers, he adopted an incisive and at times sarcastic style. The first of the writings below is a report of astronomical discoveries, written in 1610 to his patron and former pupil, Cosimo de'Medici. The second writing is Galileo's eloquent defense of his discoveries against the ostensibly religious attacks of his opponents. Published in 1615, it was dedicated to the Grand Duchess of Tuscany.

THE
STARRY MESSENGER

Revealing great, unusual, and remarkable spectacles, opening these
to the consideration of every man,
and especially of philosophers and
astronomers;

AS OBSERVED BY GALILEO GALILEI
Gentleman of Florence
Professor of Mathematics in the
University of Padua,

WITH THE AID OF A
SPYGLASS
lately invented by him,
In the surface of the Moon, in innumerable
Fixed Stars, in Nebulae, and above all
in FOUR PLANETS
swiftly revolving about Jupiter at
differing distances and periods,
and known to no one before the
Author recently perceived them
and decided that they should
be named
THE MEDICEAN STARS

Venice
1610

To the Most Serene
Cosimo II de' Medici
Fourth Grand Duke of Tuscany

Surely a distinguished public service has been rendered by those who have protected from envy the noble achievements of men who have excelled in virtue, and have thus preserved from oblivion and neglect those names which deserve immortality. In this way images sculptured in marble or cast in bronze have been handed down to posterity; to this we owe our statues, both pedestrian and equestrian; thus have we those columns and pyramids whose expense (as the poet says) reaches to the stars; finally, thus cities have been built to bear the names of men deemed worthy by posterity of commendation to all the ages. For the nature of the human mind is such that unless it is stimulated by images of things acting upon it from without, all remembrance of them passes easily away.

Looking to things even more stable and enduring, others have entrusted the immortal fame of illustrious men not to marble and metal but to the custody of the Muses and to imperishable literary monuments. But why dwell upon these things as though human wit were satisfied with earthly regions and had not dared advance beyond? For, seeking further, and well understanding that all human monuments ultimately perish through the violence of the elements or by old age, ingenuity has in fact found still more incorruptible monuments over which voracious time and envious age have been unable to assert any rights. Thus turning to the sky, man's wit has inscribed on the familiar and everlasting orbs of most bright stars the names of those whose eminent and godlike deeds have caused them to be accounted worthy of eternity in the company of the stars. And so the fame of Jupiter, of Mars, of Mercury, Hercules, and other heroes by whose names the stars are called, will not fade before the extinction of the stars themselves.

Yet this invention of human ingenuity, noble and admirable as it is, has for many centuries been out of style. Primeval heroes are in possession of those bright abodes, and hold them in their own right. In vain did the piety of Augustus attempt to elect Julius Caesar into their number, for when he tried to give the name of "Julian" to a star which appeared in his time (one of those bodies which the Greeks call "comets" and which the Romans likewise named for their hairy appearance), it vanished in a brief time and mocked his too ambitious wish. But we are able, most serene Prince, to read Your Highness in the heavens far more accurately and auspiciously. For scarce have the immortal graces of your spirit begun to shine on earth when in the heavens bright stars appear as tongues to tell and celebrate your exceeding virtues to all time. Behold, then, four stars reserved to bear your famous name; bodies which belong not to the inconspicuous multitude of fixed stars, but to the bright ranks of the planets. Variously moving about most noble Jupiter as children of his own, they complete their orbits with marvelous velocity—at the same time executing with one harmonious accord mighty revolutions every dozen years about the center of the universe; that is, the sun.[2]

Indeed, the Maker of the stars himself has seemed by clear indications to direct that I assign to these new planets Your Highness's famous name in preference to all others. For just as these stars, like children worthy of their sire, never leave the side of Jupiter by any appreciable distance, so (as indeed who does not know?) clemency, kindness of

[2] This is the first published intimation by Galileo that he accepted the Copernican system. Tycho had made Jupiter revolve about the sun, but considered the earth to be the center of the universe. It was not until 1613, however, that Galileo unequivocally supported Copernicus in print.

ASTRONOMICAL MESSAGE
Which contains and explains recent observations
made with the aid of a new spyglass[3]
concerning the surface of the moon,
the Milky Way, nebulous stars, and
innumerable fixed stars,
as well as four planets never before seen, and
now named
THE MEDICEAN STARS

Great indeed are the things which in this brief treatise I propose for observation and consideration by all students of nature. I say great, because of the excellence of the subject itself, the entirely unexpected and novel character of these things, and finally because of the instrument by means of which they have been revealed to our senses.

Surely it is a great thing to increase the numerous host of fixed stars previously visible to the unaided vision, adding countless more which have never before been seen, exposing these plainly to the eye in numbers ten times exceeding the old and familiar stars.

It is a very beautiful thing, and most gratifying to the sight, to behold the body of the moon, distant from us almost sixty earthly radii,[4] as if it were no farther away than

[3] The word "telescope" was not coined until 1611. A detailed account of its origin is given by Edward Rosen in *The Naming of the Telescope* (New York, 1947). In the present translation the modern term has been introduced for the sake of dignity and ease of reading, but only after the passage in which Galileo describes the circumstances which led him to construct the instrument (pp. 28–29).

[4] The original text reads "diameters" here and in another place. That this error was Galileo's and not the printer's has been convincingly shown by Edward Rosen (*Isis*, 1952, pp.

two such measures—so that its diameter appears almost thirty times larger, its surface nearly nine hundred times, and its volume twenty-seven thousand times as large as when viewed with the naked eye. In this way one may learn with all the certainty of sense evidence that the moon is not robed in a smooth and polished surface but is in fact rough and uneven, covered everywhere, just like the earth's surface, with huge prominences, deep valleys, and chasms.

Again, it seems to me a matter of no small importance to have ended the dispute about the Milky Way by making its nature manifest to the very senses as well as to the intellect. Similarly it will be a pleasant and elegant thing to demonstrate that the nature of those stars which astronomers have previously called "nebulous" is far different from what has been believed hitherto. But what surpasses all wonders by far, and what particularly moves us to seek the attention of all astronomers and philosophers, is the discovery of four wandering stars not known or observed by any man before us. Like Venus and Mercury, which have their own periods about the sun, these have theirs about a certain star that is conspicuous among those already known, which they sometimes precede and sometimes follow, without ever departing from it beyond certain limits. All these facts were discovered and observed by me not many days ago with the aid of a spyglass which I devised, after first being illuminated by divine grace. Perhaps other things, still more remarkable, will in time be discovered by me or by other observers with the aid of such an instrument, the form and construction of which I shall first briefly explain, as well as the occasion of its having been devised. Afterwards I shall relate the story of the observations I have made.

About ten months ago a report reached my ears that a

---

344 ff.). The slip was a curious one, as astronomers of all schools had long agreed that the maximum distance of the moon was approximately sixty terrestrial radii. Still more curious is the fact that neither Kepler nor any other correspondent appears to have called Galileo's attention to this error; not even a friend who ventured to criticize the calculations in this very passage.

certain Fleming[5] had constructed a spyglass by means of which visible objects, though very distant from the eye of the observer, were distinctly seen as if nearby. Of this truly remarkable effect several experiences were related, to which some persons gave credence while others denied them. A few days later the report was confirmed to me in a letter from a noble Frenchman at Paris, Jacques Badovere,[6] which caused me to apply myself wholeheartedly to inquire into the means by which I might arrive at the invention of a similar instrument. This I did shortly afterwards, my basis being the theory of refraction. First I prepared a tube of lead, at the ends of which I fitted two glass lenses, both plane on one side while on the other side one was spherically convex and the other concave. Then placing my eye near the concave lens I perceived objects satisfactorily large and near, for they appeared three times closer and nine times larger than when seen with the naked eye alone. Next I constructed another one, more accurate, which represented objects as enlarged more than sixty times. Finally, sparing neither labor nor expense, I succeeded in constructing for myself so excellent an instrument that objects seen by means of it appeared nearly one thousand times larger and over thirty times closer than when regarded with our natural vision.

It would be superfluous to enumerate the number and importance of the advantages of such an instrument at sea as well as on land. But forsaking terrestrial observations, I turned to celestial ones, and first I saw the moon from as near at hand as if it were scarcely two terrestrial radii away. After that I observed often with wondering delight both the planets and the fixed stars, and since I saw these latter to be very crowded, I began to seek (and eventually found)

[5] Credit for the original invention is generally assigned to Hans Lipperhey, a lens grinder in Holland who chanced upon this property of combined lenses and applied for a patent on it in 1608.

[6] Badovere studied in Italy toward the close of the sixteenth century and is said to have been a pupil of Galileo's about 1598. When he wrote concerning the new instrument in 1609 he was in the French diplomatic service at Paris, where he died in 1620.

a method by which I might measure their distances apart.

Here it is appropriate to convey certain cautions to all who intend to undertake observations of this sort, for in the first place it is necessary to prepare quite a perfect telescope, which will show all objects bright, distinct, and free from any haziness, while magnifying them at least four hundred times and thus showing them twenty times closer. Unless the instrument is of this kind it will be vain to attempt to observe all the things which I have seen in the heavens, and which will presently be set forth. Now in order to determine without much trouble the magnifying power of an instrument, trace on paper the contour of two circles or two squares of which one is four hundred times as large as the other, as it will be when the diameter of one is twenty times that of the other. Then, with both these figures attached to the same wall, observe them simultaneously from a distance, looking at the smaller one through the telescope and at the larger one with the other eye unaided. This may be done without inconvenience while holding both eyes open at the same time; the two figures will appear to be of the same size if the instrument magnifies objects in the desired proportion.

Such an instrument having been prepared, we seek a method of measuring distances apart. This we shall accomplish by the following contrivance.

Let ABCD be the tube and E be the eye of the observer. Then if there were no lenses in the tube, the rays would reach the object FG along the straight lines ECF and EDG. But when the lenses have been inserted, the rays go along the refracted lines ECH and EDI; thus they are brought closer together, and those which were previously directed freely to the object FG now include only the portion of it HI. The ratio of the distance EH to the line HI then being heart, gentleness of manner, splendor of royal blood, nobility in public affairs, and excellency of authority and rule have all fixed their abode and habitation in Your Highness. And who, I ask once more, does not know that all these virtues emanate from the benign star of Jupiter, next after God as the source of all things good? Jupiter; Jupiter, I say, at the instant of Your Highness's birth, having already emerged from the turbid mists of the horizon and occupied the midst of the heavens, illuminating the eastern sky from his own royal house, looked out from that exalted throne upon your auspicious birth and poured forth all his splendor and majesty in order that your tender body and your mind (already adorned by God with the most noble ornaments) might imbibe with their first breath that universal influence and power.

But why should I employ mere plausible arguments, when I may prove my conclusion absolutely? It pleased Almighty God that I should instruct Your Highness in mathematics, which I did four years ago at that time of year when it is customary to rest from the most exacting studies. And since clearly it was mine by divine will to serve Your Highness and thus to receive from near at hand the rays of your surpassing clemency and beneficence, what wonder is it that my heart is so inflamed as to think both day and night of little else than how I, who am indeed your subject not only by choice but by birth and lineage, may become known to you as most grateful and most anxious for your glory? And so, most serene Cosimo, having discovered under your patronage these stars unknown to every astronomer before me, I have with good right decided to designate them by the august name of your family. And if I am first to have investigated them, who can justly blame me if I likewise name them, calling them the Medicean Stars, in the hope that this name will bring as much honor to them as the names of other heroes have bestowed on other stars? For, to say nothing of Your Highness's most serene ancestors, whose everlasting glory is testified by the monuments of all history, your virtue alone, most worthy Sire, can confer upon these stars an immortal name. No one can doubt that you will fulfill those expectations, high though they are, which you have aroused by the auspicious beginning of your reign, and will not only meet but far surpass them. Thus when you have conquered your equals you may still vie with yourself, and you and your greatness will become greater every day.

Accept then, most clement Prince, this gentle glory reserved by the stars for you. May you long enjoy those blessings which are sent to you not so much from the stars as from God, their Maker and their Governor.

Your Highness's most devoted servant,

GALILEO GALILEI

PADUA, March 12, 1610

137

found, one may by means of a table of sines determine the size of the angle formed at the eye by the object HI, which we shall find to be but a few minutes of arc. Now, if to the lens CD we fit thin plates, some pierced with larger and some with smaller apertures, putting now one plate and now another over the lens as required, we may form at pleasure different angles subtending more or fewer minutes of arc, and by this means we may easily measure the intervals between stars which are but a few minutes apart, with no greater error than one or two minutes. And for the present let it suffice that we have touched lightly on these matters and scarcely more than mentioned them, as on some other occasion we shall explain the entire theory of this instrument.

Now let us review the observations made during the past two months, once more inviting the attention of all who are eager for true philosophy to the first steps of such important contemplations. Let us speak first of that surface of the moon which faces us. For greater clarity I distinguish two parts of this surface, a lighter and a darker; the lighter part seems to surround and to pervade the whole hemisphere, while the darker part discolors the moon's surface like a kind of cloud, and makes it appear covered with spots. Now those spots which are fairly dark and rather large are plain to everyone and have been seen throughout the ages; these I shall call the "large" or "ancient" spots, distinguishing them from others that are smaller in size but so numerous as to occur all over the lunar surface, and especially the lighter part. The latter spots had never been seen by anyone before me. From observations of these spots repeated many times I have been led to the opinion and conviction that the surface of the moon is not smooth, uniform, and precisely spherical as a great number of philosophers believe it (and the other heavenly bodies) to be, but is uneven, rough, and full of cavities and prominences, being not unlike the face of the earth, relieved by chains of mountains and deep valleys. The things I have seen by which I was enabled to draw this conclusion are as follows.

On the fourth or fifth day after new moon, when the moon is seen with brilliant horns, the boundary which divides the dark part from the light does not extend uniformly in an oval line as would happen on a perfectly spherical solid, but traces out an uneven, rough, and very wavy line as shown in the figure below. Indeed, many luminous excrescences extend beyond the boundary into the darker portion, while on the other hand some dark patches invade the illuminated part. Moreover a great quantity of small blackish spots, entirely separated from the dark region, are scattered almost all over the area illuminated by the sun with the exception only of that part which is occupied by the large and ancient spots. Let us note, however, that the said small spots always agree in having their blackened parts directed toward the sun, while on the side opposite the sun they are crowned with bright contours, like shining summits. There is a similar sight on earth about sunrise, when we behold the valleys not yet flooded with light though the mountains surrounding them are already ablaze with glowing splendor on the side opposite the sun. And just as the shadows in the hollows on earth diminish in size as the sun rises higher, so these spots on the moon lose their blackness as the illuminated region grows larger and larger.

Again, not only are the boundaries of shadow and light in the moon seen to be uneven and wavy, but still more astonishingly many bright points appear within the darkened portion of the moon, completely divided and separated from the illuminated part and at a considerable distance from it. After a time these gradually increase in size and brightness, and an hour or two later they become joined with the rest of the lighted part which has now increased in size. Meanwhile more and more peaks shoot up as if sprouting now here, now there, lighting up within the shadowed portion; these become larger, and finally they too are united with that same luminous surface which extends ever further. An illustration of this is to be seen in the figure above. And on the earth, before the rising of the sun, are not the highest peaks of the mountains illuminated by the sun's rays while the plains remain in shadow? Does not the light go on spreading while the larger central parts of those mountains are becoming illuminated? And when the sun has finally risen, does not the illumination of plains and hills finally become one? But on the moon the variety of elevations and depressions appears to surpass in every way the roughness of the terrestrial surface, as we shall demonstrate further on.

At present I cannot pass over in silence something worthy of consideration which I observed when the moon was approaching first quarter, as shown in the previous figure. Into the luminous part there extended a great dark gulf in the neighborhood of the lower cusp. When I had observed it for a long time and had seen it completely dark, a bright peak began to emerge, a little below its center, after about two hours. Gradually growing, this presented itself in a triangular shape, remaining completely detached and separated from the lighted surface. Around it three other small points soon began to shine, and finally, when the moon was about to set, this triangular shape (which had meanwhile become more widely extended) joined with the rest of the illuminated region and suddenly burst into the gulf of shadow like a vast promontory of light, surrounded still by the three bright peaks already mentioned. Beyond the ends of the cusps, both above and below, certain bright points emerged which were quite detached from the remaining lighted part, as may be seen depicted in the same figure. There were also a great number of dark spots in both the horns, especially in the lower one; those nearest the boundary of light and shadow appeared larger and darker, while those more distant from the boundary were not so dark and distinct. But in all cases, as we have mentioned earlier, the blackish portion of each spot is turned toward the source of the sun's radiance, while a bright rim surrounds the spot on the side away from the sun in the direction of the shadowy region of the moon. This part of the moon's surface, where it is spotted as the tail of a peacock is sprinkled with azure eyes, resembles those glass vases which have been plunged while still hot into cold water and have thus acquired a crackled and wavy surface, from which they receive their common name of "ice-cups."

As to the large lunar spots, these are not seen to be broken in the above manner and full of cavities and prominences; rather, they are even and uniform, and brighter patches crop up only here and there. Hence if anyone wished to revive the old Pythagorean[7] opinion that the moon is like another earth, its brighter part might very fitly represent the surface of the land and its darker region that of the water. I have never doubted that if our globe were seen from afar when flooded with sunlight, the land regions would appear brighter and the watery regions darker.[8] The large spots in the moon are also seen to be less elevated

---

[7] Pythagoras was a mathematician and philosopher of the sixth century B.C., a semilegendary figure whose followers were credited at Galileo's time with having anticipated the Copernican system. This tradition was based upon a misunderstanding. The Pythagoreans made the earth revolve about a "central fire" whose light and heat were reflected to the earth by the sun.

[8] Leonardo da Vinci had previously suggested that the dark and light regions of the moon were bodies of land and water, though Galileo probably did not know this. Da Vinci, however, had mistakenly supposed that the water would appear brighter than the land.

than the brighter tracts, for whether the moon is waxing or waning there are always seen, here and there along its boundary of light and shadow, certain ridges of brighter hue around the large spots (and we have attended to this in preparing the diagrams); the edges of these spots are not only lower, but also more uniform, being uninterrupted by peaks or ruggedness.

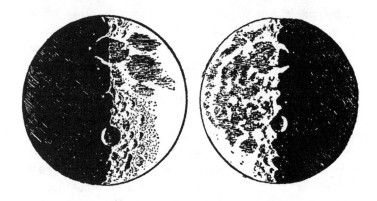

Near the large spots the brighter part stands out particularly in such a way that before first quarter and toward last quarter, in the vicinity of a certain spot in the upper (or northern) region of the moon, some vast prominences arise both above and below as shown in the figures reproduced below. Before last quarter this same spot is seen to be walled about with certain blacker contours which, like the loftiest mountaintops, appear darker on the side away from the sun and brighter on that which faces the sun. (This is the opposite of what happens in the cavities, for there the part away from the sun appears brilliant, while that which is turned toward the sun is dark and in shadow.) After a time, when the lighted portion of the moon's surface has diminished in size and when all (or nearly all) the said spot is covered with shadow, the brighter ridges of the mountains gradually emerge from the shade. This double aspect of the spot is illustrated in the ensuing figures.

There is another thing which I must not omit, for I beheld it not without a certain wonder; this is that almost in the center of the moon there is a cavity larger than all the rest, and perfectly round in shape. I have observed it near both first and last quarters, and have tried to represent it as correctly as possible in the second of the above figures. As to light and shade, it offers the same appearance as would a region like Bohemia[9] if that were enclosed on all sides by very lofty mountains arranged exactly in a circle. Indeed, this area on the moon is surrounded by such enormous peaks that the bounding edge adjacent to the dark portion of the moon is seen to be bathed in sunlight before the boundary of light and shadow reaches halfway across the same space. As in other spots, its shaded portion faces the sun while its lighted part is toward the dark side of the moon; and for a third time I draw attention to this as a very cogent proof of the ruggedness and unevenness that pervades all the bright region of the moon. Of these spots, moreover, those are always darkest which touch the boundary line between light and shadow, while those farther off

[9] This casual comparison between a part of the moon and a specific region on earth was later the basis of much trouble for Galileo; see the letter of G. Ciampoli, p. 158. Even in antiquity the idea that the moon (or any other heavenly body) was of the same nature as the earth had been dangerous to hold. The Athenians banished the philosopher Anaxagoras for teaching such notions, and charged Socrates with blasphemy for repeating them.

appear both smaller and less dark, so that when the moon ultimately becomes full (at opposition[10] to the sun), the shade of the cavities is distinguished from the light of the places in relief by a subdued and very tenuous separation.

The things we have reviewed are to be seen in the brighter region of the moon. In the large spots, no such contrast of depressions and prominences is perceived as that which we are compelled to recognize in the brighter parts by the changes of aspect that occur under varying illumination by the sun's rays throughout the multiplicity of positions from which the latter reach the moon. In the large spots there exist some holes rather darker than the rest, as we have shown in the illustrations. Yet these present always the same appearance, and their darkness is neither intensified nor diminished, although with some minute difference they appear sometimes a little more shaded and sometimes a little lighter according as the rays of the sun fall on them more or less obliquely. Moreover, they join with the neighboring regions of the spots in a gentle linkage, the boundaries mixing and mingling. It is quite different with the spots which occupy the brighter surface of the moon; these, like precipitous crags having rough and jagged peaks, stand out starkly in sharp contrasts of light and shade. And inside the large spots there are observed certain other zones that are brighter, some of them very bright indeed. Still, both these and the darker parts present always the same appearance; there is no change either of shape or of light and shadow; hence one may affirm beyond any doubt that they owe their appearance to some real dissimilarity of parts. They cannot be attributed merely to irregularity of shape, wherein shadows move in consequence of varied illuminations from the sun, as indeed is the case with the other, smaller, spots which occupy the brighter part of the moon and which change, grow, shrink,

[10] Opposition of the sun and moon occurs when they are in line with the earth between them (full moon, or lunar eclipse); conjunction, when they are in line on the same side of the earth (new moon, or eclipse of the sun).

or disappear from one day to the next, as owing their origin only to shadows of prominences.

But here I foresee that many persons will be assailed by uncertainty and drawn into a grave difficulty, feeling constrained to doubt a conclusion already explained and confirmed by many phenomena. If that part of the lunar surface which reflects sunlight more brightly is full of chasms (that is, of countless prominences and hollows), why is it that the western edge of the waxing moon, the eastern edge of the waning moon, and the entire periphery of the full moon are not seen to be uneven, rough, and wavy? On the contrary they look as precisely round as if they were drawn with a compass; and yet the whole periphery consists of that brighter lunar substance which we have declared to be filled with heights and chasms. In fact not a single one of the great spots extends to the extreme periphery of the moon, but all are grouped together at a distance from the edge.

Now let me explain the twofold reason for this troublesome fact, and in turn give a double solution to the difficulty. In the first place, if the protuberances and cavities in the lunar body existed only along the extreme edge of the circular periphery bounding the visible hemisphere, the moon might (indeed, would necessarily) look to us almost like a toothed wheel, terminated by a warty or wavy edge. Imagine, however, that there is not a single series of prominences arranged only along the very circumference, but a great many ranges of mountains together with their valleys and canyons disposed in ranks near the edge of the moon, and not only in the hemisphere visible to us but everywhere near the boundary line of the two hemispheres. Then an eye viewing them from afar will not be able to detect the separation of prominences by cavities, because the intervals between the mountains located in a given circle or a given chain will be hidden by the interposition of other heights situated in yet other ranges. This will be especially true if the eye of the observer is placed in the same straight line with the summits of these elevations. Thus on earth the summits of several mountains close together appear to be

145

situated in one plane if the spectator is a long way off and is placed at an equal elevation. Similarly in a rough sea the tops of the waves seem to lie in one plane, though between one high crest and another there are many gulfs and chasms of such depth as not only to hide the hulls but even the bulwarks, masts, and rigging of stately ships. Now since there are many chains of mountains and chasms on the moon in addition to those around its periphery, and since the eye, regarding these from a great distance, lies nearly in the plane of their summits, no one need wonder that they appear as arranged in a regular and unbroken line.

To the above explanation another may be added; namely, that there exists around the body of the moon, just as around the earth, a globe of some substance denser than the rest of the aether.[11] This may serve to receive and reflect the sun's radiations without being sufficiently opaque to prevent our seeing through it, especially when it is not il-luminated. Such a globe, lighted by the sun's rays, makes the body of the moon appear larger than it really is, and if it were thicker it would be able to prevent our seeing the actual body of the moon. And it actually is thicker near the circumference of the moon; I do not mean in an absolute sense, but relatively to the rays of our vision, which cut it obliquely there. Thus it may obstruct our vision, especially when it is lighted, and cloak the lunar periphery that is exposed to the sun. This may be more clearly understood from the figure below, in which the body of the moon, ABC, is surrounded by the vaporous globe DEG.

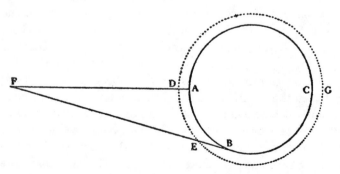

[11] The aether, or "ever-moving," was the special substance of

The eyesight from F reaches the moon in the central region, at A for example, through a lesser thickness of the vapors DA, while toward the extreme edges a deeper stratum of vapors, EB, limits and shuts out our sight. One indication of this is that the illuminated portion of the moon appears to be larger in circumference than the rest of the orb, which lies in shadow. And perhaps this same cause will appeal to some as reasonably explaining why the larger spots on the moon are nowhere seen to reach the very edge, probable though it is that some should occur there. Possibly they are invisible by being hidden under a thicker and more luminous mass of vapors.

That the lighter surface of the moon is everywhere dotted with protuberances and gaps has, I think, been made sufficiently clear from the appearances already explained. It remains for me to speak of their dimensions, and to show that the earth's irregularities are far less than those of the moon. I mean that they are absolutely less, and not merely in relation to the sizes of the respective globes. This is plainly demonstrated as follows.

I had often observed, in various situations of the moon with respect to the sun, that some summits within the shadowy portion appeared lighted, though lying some distance from the boundary of the light. By comparing this separation to the whole diameter of the moon, I found that it sometimes exceeded one-twentieth of the diameter. Accordingly, let CAF be a great circle of the lunar body, E its center, and CF a diameter, which is to the diameter of the earth as two is to seven.

Since according to very precise observations the diameter of the earth is seven thousand miles, CF will be two thousand, CE one thousand, and one-twentieth of CF will be one hundred miles. Now let CF be the diameter of the great circle which divides the light part of the moon from the dark part (for because of the very great distance of the

---

which the sky and all the heavenly bodies were supposed to be made, a substance essentially different from all the earthly "elements." In later years Galileo abandoned his suggestion here that the moon has a vaporous atmosphere.

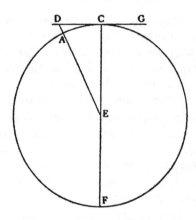

sun from the moon, this does not differ appreciably from a great circle), and let A be distant from C by one-twentieth of this. Draw the radius EA, which, when produced, cuts the tangent line GCD (representing the illuminating ray) in the point D. Then the arc CA, or rather the straight line CD, will consist of one hundred units whereof CE contains one thousand, and the sum of the squares of DC and CE will be 1,010,000. This is equal to the square of DE; hence ED will exceed 1,004, and AD will be more than four of those units of which CE contains one thousand. Therefore the altitude AD on the moon, which represents a summit reaching up to the solar ray GCD and standing at the distance CD from C, exceeds four miles. But on the earth we have no mountains which reach to a perpendicular height of even one mile.[12] Hence it is quite clear that the prominences on the moon are loftier than those on the earth.

Here I wish to assign the cause of another lunar phenomenon well worthy of notice. I observed this not just recently, but many years ago, and pointed it out to some of my friends and pupils, explaining it to them and giving its

[12] Galileo's estimate of four miles for the height of some lunar mountains was a very good one. His remark about the maximum height of mountains on the earth was, however, quite mistaken. An English propagandist for his views, John Wilkins, took pains to correct this error in his anonymous *Discovery of a New World . . . in the Moon* (London, 1638), Prop. ix.

true cause. Yet since it is rendered more evident and easier to observe with the aid of the telescope, I think it not unsuitable for introduction in this place, especially as it shows more clearly the connection between the moon and the earth.

When the moon is not far from the sun, just before or after new moon, its globe offers itself to view not only on the side where it is adorned with shining horns, but a certain faint light is also seen to mark out the periphery of the dark part which faces away from the sun, separating this from the darker background of the aether. Now if we examine the matter more closely, we shall see that not only does the extreme limb of the shaded side glow with this uncertain light, but the entire face of the moon (including the side which does not receive the glare of the sun) is whitened by a not inconsiderable gleam. At first glance only a thin luminous circumference appears, contrasting with the darker sky coterminous with it; the rest of the surface appears darker from its contact with the shining horns which distract our vision. But if we place ourselves so as to interpose a roof or chimney or some other object at a considerable distance from the eye, the shining horns may be hidden while the rest of the lunar globe remains exposed to view. It is then found that this region of the moon, though deprived of sunlight, also shines not a little. The effect is heightened if the gloom of night has already deepened through departure of the sun, for in a darker field a given light appears brighter.

Moreover, it is found that this secondary light of the moon (so to speak) is greater according as the moon is closer to the sun. It diminishes more and more as the moon recedes from that body until, after the first quarter and before the last, it is seen very weakly and uncertainly even when observed in the darkest sky. But when the moon is within sixty degrees of the sun it shines remarkably, even in twilight; so brightly indeed that with the aid of a good telescope one may distinguish the large spots. This remarkable gleam has afforded no small perplexity to philosophers, and in order to assign a cause for it some have offered one idea and some another. Some would say it is an inherent

149

and natural light of the moon's own; others, that it is imparted by Venus; others yet, by all the stars together; and still others deiive it from the sun, whose rays they would have permeate the thick solidity of the moon. But statements of this sort are refuted and their falsity evinced with little difficulty. For if this kind of light were the moon's own, or were contributed by the stars, the moon would retain it and would display it particularly during eclipses, when it is left in an unusually dark sky. This is contradicted by experience, for the brightness which is seen on the moon during eclipses is much fainter and is ruddy, almost copper-colored, while this is brighter and whitish. Moreover the other light is variable and movable, for it covers the face of the moon in such a way that the place near the edge of the earth's shadow is always seen to be brighter than the rest of the moon; this undoubtedly results from contact of the tangent solar rays with some denser zone which girds the moon about.[13] By this contact a sort of twilight is diffused over the neighboring regions of the moon, just as on earth a sort of crepuscular light is spread both morning and evening; but with this I shall deal more fully in my book on the system of the world.[14]

To assert that the moon's secondary light is imparted by Venus is so childish as to deserve no reply. Who is so ignorant as not to understand that from new moon to a separation of sixty degrees between moon and sun, no part of the moon which is averted from the sun can possibly be seen from Venus? And it is likewise unthinkable that this light should depend upon the sun's rays penetrating the thick

[13] Kepler had correctly accounted for the existence of this light and its ruddy color. It is caused by refraction of sunlight in the earth's atmosphere, and does not require a lunar atmosphere as supposed by Galileo.

[14] The book thus promised was destined not to appear for more than two decades. Events which will presently be recounted prevented its publication for many years, and then it had to be modified to present the arguments for both the Ptolemaic and Copernican systems instead of just the latter as Galileo here planned. Even then it was suppressed, and the author was condemned to life imprisonment.

solid mass of the moon, for then this light would never dwindle, inasmuch as one hemisphere of the moon is always illuminated except during lunar eclipses. And the light does diminish as the moon approaches first quarter, becoming completely obscured after that is passed.

Now since the secondary light does not inherently belong to the moon, and is not received from any star or from the sun, and since in the whole universe there is no other body left but the earth, what must we conclude? What is to be proposed? Surely we must assert that the lunar body (or any other dark and sunless orb) is illuminated by the earth. Yet what is there so remarkable about this? The earth, in fair and grateful exchange, pays back to the moon an illumination similar to that which it receives from her throughout nearly all the darkest gloom of night.

Let us explain this matter more fully. At conjunction the moon occupies a position between the sun and the earth; it is then illuminated by the sun's rays on the side which is turned away from the earth. The other hemisphere, which faces the earth, is covered with darkness; hence the moon does not illuminate the surface of the earth at all. Next, departing gradually from the sun, the moon comes to be lighted partly upon the side it turns toward us, and its whitish horns, still very thin, illuminate the earth with a faint light. The sun's illumination of the moon increasing now as the moon approaches first quarter, a reflection of that light to the earth also increases. Soon the splendor on the moon extends into a semicircle, and our nights grow brighter; at length the entire visible face of the moon is irradiated by the sun's resplendent rays, and at full moon the whole surface of the earth shines in a flood of moonlight. Now the moon, waning, sends us her beams more weakly, and the earth is less strongly lighted; at length the moon returns to conjunction with the sun, and black night covers the earth.

In this monthly period, then, the moonlight gives us alternations of brighter and fainter illumination; and the benefit is repaid by the earth in equal measure. For while the moon is between us and the sun (at new moon), there lies before it the entire surface of that hemisphere of the earth which is expos d to the sun and illuminated by vivid rays. The moon rece ves the light which this reflects, and thus the nearer hemisphere of the moon—that is, the one deprived of sunlight—appears by virtue of this illumination to be not a little luminous. When the moon is ninety degrees away from the sun it sees but half the earth illuminated (the western half), for the other (the eastern half) is enveloped in night. Hence the moon itself is illuminated less brightly from the earth, and as a result its secondary light appears fainter to us. When the moon is in opposition to the sun, it faces a hemisphere of the earth that is steeped in the gloom of night, and if this position occurs in the plane of the ecliptic the moon will receive no light at all, being deprived of both the solar and the terrestrial rays. In its various other positions with respect to the earth and sun, the moon receives more or less light according as it faces a greater or smaller portion of the illuminated hemisphere of the earth. And between these two globes a relation is maintained such that whenever the earth is most brightly lighted by the moon, the moon is least lighted by the earth, and vice versa.

Let these few remarks suffice us here concerning this matter, which will be more fully treated in our *System of the world*. In that book, by a multitude of arguments and experiences, the solar reflection from the earth will be shown to be quite real—against those who argue that the earth must be excluded from the dancing whirl of stars for the specific reason that it is devoid of motion and of light. We shall prove the earth to be a wandering body surpassing the moon in splendor, and not the sink of all dull refuse of the universe; this we shall support by an infinitude of arguments drawn from nature.

Thus far we have spoken of our observations concerning the body of the moon. Let us now set forth briefly what has thus far been observed regarding the fixed stars. And first of all, the following fact deserves consideration: The stars, whether fixed or wandering,[15] appear not to be enlarged by the telescope in the same proportion as that in which it magnifies other objects, and even the moon itself. In the stars this enlargement seems to be so much less that a telescope which is sufficiently powerful to magnify other objects a hundredfold is scarcely able to enlarge the stars four or five times. The reason for this is as follows.

When stars are viewed by means of unaided natural vision, they present themselves to us not as of their simple (and, so to speak, their physical) size, but as irradiated by a certain fulgor and as fringed with sparkling rays, especially when the night is far advanced. From this they appear larger than they would if stripped of those adventitious hairs of light, for the angle at the eye is determined not by the primary body of the star but by the brightness which extends so widely about it. This appears quite clearly from the fact that when stars first emerge from twilight at sunset they look very small, even if they are of the first magnitude; Venus itself, when visible in broad daylight, is so small as scarcely to appear equal to a star of the sixth magnitude. Things fall out differently with other objects, and even with the moon itself; these, whether seen in daylight or the deepest night, appear always of the same bulk. Therefore the stars are seen crowned among shadows, while daylight is able to remove their headgear; and not daylight alone, but any thin cloud that interposes itself between a star and the eye of the observer. The same effect is produced by black veils or colored glasses, through the interposition of which obstacles the stars are abandoned by their surrounding brilliance. A telescope similarly accomplishes the same result. It removes from the stars their adventitious and accidental rays, and then it enlarges their simple globes (if indeed the stars are naturally globular) so that they seem to be magnified in a lesser ratio than other objects. In fact a star of the fifth or sixth magnitude when seen through a telescope presents itself as one of the first magnitude.

[15] That is, planets. Among these bodies Galileo counted his newly discovered satellites of Jupiter. The term "satellites" was introduced somewhat later by Kepler.

Deserving of notice also is the difference between the appearances of the planets and of the fixed stars.[16] The planets show their globes perfectly round and definitely bounded, looking like little moons, spherical and flooded all over with light; the fixed stars are never seen to be bounded by a circular periphery, but have rather the aspect of blazes whose rays vibrate about them and scintillate a great deal. Viewed with a telescope they appear of a shape similar to that which they present to the naked eye, but sufficiently enlarged so that a star of the fifth or sixth magnitude seems to equal the Dog Star, largest of all the fixed stars. Now, in addition to stars of the sixth magnitude, a host of other stars are perceived through the telescope which escape the naked eye; these are so numerous as almost to surpass belief. One may, in fact, see more of them than all the stars included among the first six magnitudes. The largest of these, which we may call stars of the seventh magnitude, or the first magnitude of invisible stars, appear through the telescope as larger and brighter than stars of the second magnitude when the latter are viewed with the naked eye. In order to give one or two proofs of their almost inconceivable number, I have adjoined pictures of two constellations. With these as samples, you may judge of all the others.

In the first I had intended to depict the entire constellation of Orion, but I was overwhelmed by the vast quantity of stars and by limitations of time, so I have deferred this to another occasion. There are more than five hundred new stars distributed among the old ones within limits of one or two degrees of arc. Hence to the three stars in the Belt of Orion and the six in the Sword which were previously known, I have added eighty adjacent stars discovered re-

[16] Fixed stars are so distant that their light reaches the earth as from dimensionless points. Hence their images are not enlarged by even the best telescopes, which serve only to gather more of their light and in that way increase their visibility. Galileo was never entirely clear about this distinction. Nevertheless, by applying his knowledge of the effects described here, he greatly reduced the prevailing overestimation of visual dimensions of stars and planets.

cently, preserving the intervals between them as exactly as I could. To distinguish the known or ancient stars, I have depicted them larger and have outlined them doubly; the other (invisible) stars I have drawn smaller and without the extra line. I have also preserved differences of magnitude as well as possible.

The Belt and Sword of Orion

In the second example I have depicted the six stars of Taurus known as the Pleiades (I say six, inasmuch as the seventh is hardly ever visible) which lie within very narrow limits in the sky. Near them are more than forty others, invisible, no one of which is much more than half a degree away from the original six. I have shown thirty-six of these in the diagram; as in the case of Orion I have preserved their intervals and magnitudes, as well as the distinction between old stars and new.

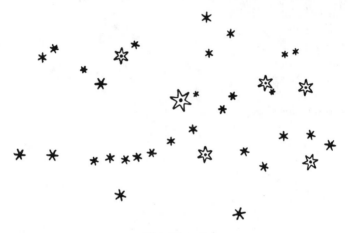

The Pleiades

Third, I have observed the nature and the material of the Milky Way. With the aid of the telescope this has been scrutinized so directly and with such ocular certainty that all the disputes which have vexed philosophers through so many ages have been resolved, and we are at last freed from wordy debates about it. The galaxy is, in fact, nothing but a congeries of innumerable stars grouped together in clusters. Upon whatever part of it the telescope is directed, a vast crowd of stars is immediately presented to view. Many of them are rather large and quite bright, while the number of smaller ones is quite beyond calculation.

But it is not only in the Milky Way that whitish clouds are seen; several patches of similar aspect shine with faint light here and there throughout the aether, and if the telescope is turned upon any of these it confronts us with a tight mass of stars. And what is even more remarkable, the stars which have been called "nebulous" by every astronomer up to this time turn out to be groups of very small stars arranged in a wonderful manner. Although each star separately escapes our sight on account of its smallness or the immense distance from us, the mingling of their rays gives rise to that gleam which was formerly believed to be some denser part of the aether that was capable of reflecting rays from stars or from the sun. I have observed some of these constellations and have decided to depict two of them.

In the first you have the nebula called the Head of Orion, in which I have counted twenty-one stars. The second contains the nebula called Praesepe,[17] which is not a single star but a mass of more than forty starlets. Of these I have shown thirty-six, in addition to the Aselli, arranged in the order shown.

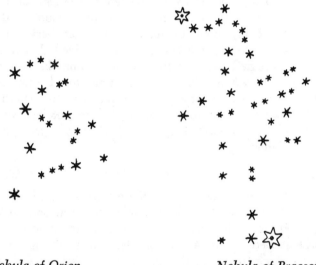

Nebula of Orion   *Nebula of Praesepe*

We have now briefly recounted the observations made thus far with regard to the moon, the fixed stars, and the Milky Way. There remains the matter which in my opinion deserves to be considered the most important of all—the disclosure of four PLANETS never seen from the creation of

[17] Praesepe, "the Manger," is a small whitish cluster of stars lying between the two Aselli (ass-colts) which are imagined as feeding from it. It lies in the constellation Cancer.

the world up to our own time, together with the occasion of my having discovered and studied them, their arrangements, and the observations made of their movements and alterations during the past two months. I invite all astronomers to apply themselves to examine them and determine their periodic times, something which has so far been quite impossible to complete, owing to the shortness of the time. Once more, however, warning is given that it will be necessary to have a very accurate telescope such as we have described at the beginning of this discourse.

On the seventh day of January in this present year 1610, at the first hour of night, when I was viewing the heavenly bodies with a telescope, Jupiter presented itself to me; and because I had prepared a very excellent instrument for myself, I perceived (as I had not before, on account of the weakness of my previous instrument) that beside the planet there were three starlets, small indeed, but very bright. Though I believed them to be among the host of fixed stars, they aroused my curiosity somewhat by appearing to lie in an exact straight line parallel to the ecliptic, and by their being more splendid than others of their size. Their arrangement with respect to Jupiter and each other was the following:

East   ✳   ✳   ◯   ✳   *West*

that is, there were two stars on the eastern side and one to the west. The most easterly star and the western one appeared larger than the other. I paid no attention to the distances between them and Jupiter, for at the outset I thought them to be fixed stars, as I have said.[18] But re-

[18] The reader should remember that the telescope was nightly revealing to Galileo hundreds of fixed stars never previously observed. His unusual gifts for astronomical observation are illustrated by his having noticed and remembered these three merely by reason of their alignment, and recalling them so well that when by chance he happened to see them the following night he was certain that they had changed their positions. No such plausible and candid account of the discovery was given by the rival astronomer Simon Mayr, who four years later claimed priority. See pp. 233 ff. and note 4, pp. 233–34.

turning to the same investigation on January eighth—led by what, I do not know—I found a very different arrangement. The three starlets were now all to the west of Jupiter, closer together, and at equal intervals from one another as shown in the following sketch:

East   ⭘ ✳ ✳ ✳   *West*

At this time, though I did not yet turn my attention to the way the stars had come together, I began to concern myself with the question how Jupiter could be east of all these stars when on the previous day it had been west of two of them. I commenced to wonder whether Jupiter was not moving eastward at that time, contrary to the computations of the astronomers, and had got in front of them by that motion.[19] Hence it was with great interest that I awaited the next night. But I was disappointed in my hopes, for the sky was then covered with clouds everywhere.

On the tenth of January, however, the stars appeared in this position with respect to Jupiter:

East   ✳   ✳   ⭘   *West*

that is, there were but two of them, both easterly, the third (as I supposed) being hidden behind Jupiter. As at first, they were in the same straight line with Jupiter and were arranged precisely in the line of the zodiac. Noticing this, and knowing that there was no way in which such alterations could be attributed to Jupiter's motion, yet being certain that these were still the same stars I had observed (in fact no other was to be found along the line of the zodiac for a long way on either side of Jupiter), my perplexity was now transformed into amazement. I was sure that the apparent changes belonged not to Jupiter but to the observed stars, and I resolved to pursue this investigation with greater care and attention.

And thus, on the eleventh of January, I saw the following disposition:

[19] See note 4, p. 12. Jupiter was at this time in "retrograde" motion; that is, the earth's motion made the planet appear to be moving westward among the fixed stars.

East   ✳ ✳   ◯   *West*

There were two stars, both to the east, the central one being three times as far from Jupiter as from the one farther east. The latter star was nearly double the size of the former, whereas on the night before they had appeared approximately equal.

I had now decided beyond all question that there existed in the heavens three stars wandering about Jupiter as do Venus and Mercury about the sun, and this became plainer than daylight from observations on similar occasions which followed. Nor were there just three such stars; four wanderers complete their revolutions about Jupiter, and of their alterations as observed more precisely later on we shall give a description here. Also I measured the distances between them by means of the telescope, using the method explained before. Moreover I recorded the times of the observations, especially when more than one was made during the same night—for the revolutions of these planets are so speedily completed that it is usually possible to take even their hourly variations.

Thus on the twelfth of January at the first hour of night I saw the stars arranged in this way:

East   ✳ *◯ ✳   West*

The most easterly star was larger than the western one, though both were easily visible and quite bright. Each was about two minutes of arc distant from Jupiter. The third star was invisible at first, but commenced to appear after two hours; it almost touched Jupiter on the east, and was quite small. All were on the same straight line directed along the ecliptic.

On the thirteenth of January four stars were seen by me for the first time, in this situation relative to Jupiter:

East   *   ◯ * ✳ *   West*

Three were westerly and one was to the east; they formed a straight line except that the middle western star departed slightly toward the north. The eastern star was two minutes of arc away from Jupiter, and the intervals of the rest from one another and from Jupiter were about one minute. All the stars appeared to be of the same magnitude, and though small were very bright, much brighter than fixed stars of the same size.[20]

.   .   .   .

On the twenty-sixth of February, midway in the first hour of night, there were only two stars:

East   *   O   *   West

One was to the east, ten minutes from Jupiter; the other to the west, six minutes away. The eastern one was somewhat smaller than the western. But at the fifth hour three stars were seen:

East   *   O *   *   West

In addition to the two already noticed, a third was discovered to the west near Jupiter; it had at first been hidden behind Jupiter and was now one minute away. The eastern one appeared farther away than before, being eleven minutes from Jupiter.

This night for the first time I wanted to observe the progress of Jupiter and its accompanying planets along the line of the zodiac in relation to some fixed star, and such a star was seen to the east, eleven minutes distant from the easterly starlet and a little removed toward the south, in the following manner:

East   *   O *   *   West

★

On the twenty-seventh of February, four minutes after the first hour, the stars appeared in this configuration:

[20] Galileo's day-by-day journal of observations continued in unbroken sequence until ten days before publication of the book, which he remained in Venice to supervise. The observations omitted here contained nothing of a novel character.

East   ✳   ✳○   ✳✳   West

★

The most easterly was ten minutes from Jupiter; the next, thirty seconds; the next to the west was two minutes thirty seconds from Jupiter, and the most westerly was one minute from that. Those nearest Jupiter appeared very small, while the end ones were plainly visible, especially the westernmost. They marked out an exactly straight line along the course of the ecliptic. The progress of these planets toward the east is seen quite clearly by reference to the fixed star mentioned, since Jupiter and its accompanying planets were closer to it, as may be seen in the figure above. At the fifth hour, the eastern star closer to Jupiter was one minute away.

At the first hour on February twenty-eighth, two stars only were seen; one easterly, distant nine minutes from Jupiter, and one to the west, two minutes away. They were easily visible and on the same straight line. The fixed star, perpendicular to this line, now fell under the eastern planet as in this figure:

East   ✳   ○   ✳   West

★

At the fifth hour a third star, two minutes east of Jupiter, was seen in this position:

East   ✳   ✳ ○ ✳   West

On the first of March, forty minutes after sunset, four stars all to the east were seen, of which the nearest to Jupiter was two minutes away, the next was one minute from this, the third two seconds from that and brighter than any of the others; from this in turn the most easterly was four minutes distant, and it was smaller than the rest. They marked out almost a straight line, but the third one counting from Jupiter was a little to the north. The fixed star formed an equilateral triangle with Jupiter and the most easterly star, as in this figure:

East   ✳   ✳ ✳ ✳   ◯   West

★

On March second, half an hour after sunset, there were three planets, two to the east and one to the west, in this configuration:

East   ✳ ✳   ◯   ✳   West

★

The most easterly was seven minutes from Jupiter and thirty seconds from its neighbor; the western one was two minutes away from Jupiter. The end stars were very bright and were larger than that in the middle, which appeared very small. The most easterly star appeared a little elevated toward the north from the straight line through the other planets and Jupiter. The fixed star previously mentioned was eight minutes from the western planet along the line drawn from it perpendicularly to the straight line through all the planets, as shown above.

I have reported these relations of Jupiter and its companions with the fixed star so that anyone may comprehend that the progress of those planets, both in longitude and latitude, agrees exactly with the movements derived from planetary tables.

Such are the observations concerning the four Medicean planets recently first discovered by me, and although from these data their periods have not yet been reconstructed in numerical form, it is legitimate at least to put in evidence some facts worthy of note. Above all, since they sometimes follow and sometimes precede Jupiter by the same intervals, and they remain within very limited distances either to east or west of Jupiter, accompanying that planet in both its retrograde and direct movements in a constant manner, no one can doubt that they complete their revolutions about Jupiter and at the same time effect all together a twelve-year period about the center of the universe. That they also revolve in unequal circles is manifestly deduced from the fact that at the greatest elongation[21] from Jupiter it is never possible to see two of these planets in conjunction, whereas in the vicinity of Jupiter they are found united two, three, and sometimes all four together. It is also observed that the revolutions are swifter in those planets which describe smaller circles about Jupiter, since the stars closest to Jupiter are usually seen to the east when on the previous day they appeared to the west, and vice versa, while the planet which traces the largest orbit appears upon accurate observation of its returns to have a semimonthly period.

Here we have a fine and elegant argument for quieting the doubts of those who, while accepting with tranquil mind the revolutions of the planets about the sun in the Copernican system, are mightily disturbed to have the moon alone revolve about the earth and accompany it in an annual rotation about the sun. Some have believed that this structure of the universe should be rejected as impossible. But now we have not just one planet rotating about another while both run through a great orbit around the sun; our own eyes show us four stars which wander around Jupiter as does the moon around the earth, while all together trace out a grand revolution about the sun in the space of twelve years.

And finally we should not omit the reason for which the Medicean stars appear sometimes to be twice as large as at other times, though their orbits about Jupiter are very restricted. We certainly cannot seek the cause in terrestrial vapors, as Jupiter and its neighboring fixed stars are not seen to change size in the least while this increase and diminution are taking place. It is quite unthinkable that the cause of variation should be their change of distance from

[21] By this is meant the greatest angular separation from Jupiter attained by any of the satellites.

the earth at perigee and apogee, since a small circular rotation could by no means produce this effect, and an oval motion (which in this case would have to be nearly straight) seems unthinkable and quite inconsistent with the appearances.[22] But I shall gladly explain what occurs to me on this matter, offering it freely to the judgment and criticism of thoughtful men. It is known that the interposition of terrestrial vapors makes the sun and moon appear large, while the fixed stars and planets are made to appear smaller. Thus the two great luminaries are seen larger when close to the horizon, while the stars appear smaller and for the most part hardly visible. Hence the stars appear very feeble by day and in twilight, though the moon does not, as we have said. Now from what has been said above, and even more from what we shall say at greater length in our *System,* it follows that not only the earth but also the moon is surrounded by an envelope of vapors, and we may apply precisely the same judgment to the rest of the planets. Hence it does not appear entirely impossible to assume that around Jupiter also there exists an envelope denser than the rest of the aether, about which the Medicean planets revolve as does the moon about the elemental sphere. Through the interposition of this envelope they appear larger when they are in perigee by the removal, or at least the attenuation, of this envelope.

Time prevents my proceeding further, but the gentle reader may expect more soon.

FINIS

[22] The marked variation in brightness of the satellites which Galileo observed may be attributed mainly to markings upon their surfaces, though this was not determined until two centuries later. The mention here of a possible oval shape of the orbits is the closest Galileo ever came to accepting Kepler's great discovery of the previous year (cf. p. 17). Even here, however, he was probably not thinking of Kepler's work but of an idea proposed by earlier astronomers for the moon and the planet Venus.

LETTER

GALILEO GALILEI
TO
THE MOST SERENE
GRAND DUCHESS MOTHER:

Some years ago, as Your Serene Highness well knows, I discovered in the heavens many things that had not been seen before our own age. The novelty of these things, as well as some consequences which followed from them in contradiction to the physical notions commonly held among academic philosophers, stirred up against me no small number of professors—as if I had placed these things in the sky with my own hands in order to upset nature and overturn the sciences. They seemed to forget that the increase of known truths stimulates the investigation, establishment, and growth of the arts; not their diminution or destruction.

Showing a greater fondness for their own opinions than for truth, they sought to deny and disprove the new things which, if they had cared to look for themselves, their own senses would have demonstrated to them. To this end they hurled various charges and published numerous writings filled with vain arguments, and they made the grave mistake of sprinkling these with passages taken from places in the Bible which they had failed to understand properly, and which were ill suited to their purposes.

These men would perhaps not have fallen into such error had they but paid attention to a most useful doctrine of St. Augustine's, relative to our making positive statements about things which are obscure and hard to understand by means of reason alone. Speaking of a certain physical conclusion about the heavenly bodies, he wrote: "Now keeping always our respect for moderation in grave piety, we ought not to believe anything inadvisedly on a dubious point, lest in favor to our error we conceive a prejudice against something that truth hereafter may reveal to be not contrary in any way to the sacred books of either the Old or the New Testament."[1]

Well, the passage of time has revealed to everyone the truths that I previously set forth; and, together with the truth of the facts, there has come to light the great difference in attitude between those who simply and dispassionately refused to admit the discoveries to be true, and those who combined with their incredulity some reckless passion of their own. Men who were well grounded in astronomical and physical science were persuaded as soon as they received my first message. There were others who denied them or remained in doubt only because of their novel and unexpected character, and because they had not yet had the opportunity to see for themselves. These men have by degrees come to be satisfied. But some, besides allegiance to their original error, possess I know not what fanciful interest in remaining hostile not so much toward the things in question as toward their discoverer. No longer being able to deny them, these men now take refuge in obstinate silence, but being more than ever exasperated by that which has pacified and quieted other men, they divert their thoughts to other fancies and seek new ways to damage me.

I should pay no more attention to them than to those who previously contradicted me—at whom I always laugh, being assured of the eventual outcome—were it not that in their new calumnies and persecutions I perceive that they do not stop at proving themselves more learned than I am (a claim which I scarcely contest), but go so far as to cast against me imputations of crimes which must be, and are, more abhorrent to me than death itself. I cannot remain satisfied merely to know that the injustice of this is recognized by those who are acquainted with these men and with me, as perhaps it is not known to others.

[1] *De Genesi ad literam,* end of bk. ii. (Citations of theological works are taken from Galileo's marginal notes, without verification.)

Persisting in their original resolve to destroy me and everything mine by any means they can think of, these men are aware of my views in astronomy and philosophy. They know that as to the arrangement of the parts of the universe, I hold the sun to be situated motionless in the center of the revolution of the celestial orbs while the earth rotates on its axis and revolves about the sun. They know also that I support this position not only by refuting the arguments of Ptolemy and Aristotle, but by producing many counterarguments; in particular, some which relate to physical effects whose causes can perhaps be assigned in no other way. In addition there are astronomical arguments derived from many things in my new celestial discoveries that plainly confute the Ptolemaic system while admirably agreeing with and confirming the contrary hypothesis. Possibly because they are disturbed by the known truth of other propositions of mine which differ from those commonly held, and therefore mistrusting their defense so long as they confine themselves to the field of philosophy, these men have resolved to fabricate a shield for their fallacies out of the mantle of pretended religion and the authority of the Bible. These they apply, with little judgment, to the refutation of arguments that they do not understand and have not even listened to.

First they have endeavored to spread the opinion that such propositions in general are contrary to the Bible and are consequently damnable and heretical. They know that it is human nature to take up causes whereby a man may oppress his neighbor, no matter how unjustly, rather than those from which a man may receive some just encouragement. Hence they have had no trouble in finding men who would preach the damnability and heresy of the new doctrine from their very pulpits with unwonted confidence, thus doing impious and inconsiderate injury not only to that doctrine and its followers but to all mathematics and mathematicians in general. Next, becoming bolder, and hoping (though vainly) that this seed which first took root in their hypocritical minds would send out branches and ascend to heaven, they began scattering rumors among the people that before long this doctrine would be condemned by the supreme authority. They know, too, that official condemnation would not only suppress the two propositions which I have mentioned, but would render damnable all other astronomical and physical statements and observations that have any necessary relation or connection with these.

In order to facilitate their designs, they seek so far as possible (at least among the common people) to make this opinion seem new and to belong to me alone. They pretend not to know that its author, or rather its restorer and confirmer, was Nicholas Copernicus; and that he was not only a Catholic, but a priest and a canon. He was in fact so esteemed by the church that when the Lateran Council under Leo X took up the correction of the church calendar, Copernicus was called to Rome from the most remote parts of Germany to undertake its reform. At that time the calendar was defective because the true measures of the year and the lunar month were not exactly known. The Bishop of Culm,[2] then superintendent of this matter, assigned Copernicus to seek more light and greater certainty concerning the celestial motions by means of constant study and labor. With Herculean toil he set his admirable mind to this task, and he made such great progress in this science and brought our knowledge of the heavenly motions to such precision that he became celebrated as an astronomer. Since that time not only has the calendar been regulated by his teachings, but tables of all the motions of the planets have been calculated as well.

Having reduced his system into six books, he published these at the instance of the Cardinal of Capua[3] and the Bishop of Culm. And since he had assumed his laborious enterprise by order of the supreme pontiff, he dedicated this book *On the celestial revolutions* to Pope Paul III. When printed, the book was accepted by the holy Church, and it has been read and studied by everyone without the

[2] Tiedmann Giese, to whom Copernicus referred in his preface as "that scholar, my good friend."
[3] Nicholas Schoenberg, spoken of by Copernicus as "celebrated in all fields of scholarship."

faintest hint of any objection ever being conceived against its doctrines. Yet now that manifest experiences and necessary proofs have shown them to be well grounded, persons exist who would strip the author of his reward without so much as looking at his book, and add the shame of having him pronounced a heretic. All this they would do merely to satisfy their personal displeasure conceived without any cause against another man, who has no interest in Copernicus beyond approving his teachings.

Now as to the false aspersions which they so unjustly seek to cast upon me, I have thought it necessary to justify myself in the eyes of all men, whose judgment in matters of religion and of reputation I must hold in great esteem. I shall therefore discourse of the particulars which these men produce to make this opinion detested and to have it condemned not merely as false but as heretical. To this end they make a shield of their hypocritical zeal for religion. They go about invoking the Bible, which they would have minister to their deceitful purposes. Contrary to the sense of the Bible and the intention of the holy Fathers, if I am not mistaken, they would extend such authorities until even in purely physical matters—where faith is not involved—they would have us altogether abandon reason and the evidence of our senses in favor of some biblical passage, though under the surface meaning of its words this passage may contain a different sense.

I hope to show that I proceed with much greater piety than they do, when I argue not against condemning this book, but against condemning it in the way they suggest—that is, without understanding it, weighing it, or so much as reading it. For Copernicus never discusses matters of religion or faith, nor does he use arguments that depend in any way upon the authority of sacred writings which he might have interpreted erroneously. He stands always upon physical conclusions pertaining to the celestial motions, and deals with them by astronomical and geometrical demonstrations, founded primarily upon sense experiences and very exact observations. He did not ignore the Bible, but he knew very well that if his doctrine were proved, then it could not contradict the Scriptures when they were rightly understood. And thus at the end of his letter of dedication, addressing the pope, he said:

"If there should chance to be any exegetes ignorant of mathematics who pretend to skill in that discipline, and dare to condemn and censure this hypothesis of mine upon the authority of some scriptural passage twisted to their purpose, I value them not, but disdain their unconsidered judgment. For it is known that Lactantius—a poor mathematician though in other respects a worthy author—writes very childishly about the shape of the earth when he scoffs at those who affirm it to be a globe. Hence it should not seem strange to the ingenious if people of that sort should in turn deride me. But mathematics is written for mathematicians, by whom, if I am not deceived, these labors of mine will be recognized as contributing something to their domain, as also to that of the Church over which Your Holiness now reigns."[4]

Such are the people who labor to persuade us that an author like Copernicus may be condemned without being read, and who produce various authorities from the Bible, from theologians, and from Church Councils to make us believe that this is not only lawful but commendable. Since I hold these to be of supreme authority, I consider it rank temerity for anyone to contradict them—when employed according to the usage of the holy Church. Yet I do not believe it is wrong to speak out when there is reason to suspect that other men wish, for some personal motive, to produce and employ such authorities for purposes quite different from the sacred intention of the holy Church.

Therefore I declare (and my sincerity will make itself manifest) not only that I mean to submit myself freely and renounce any errors into which I may fall in this discourse through ignorance of matters pertaining to religion, but that I do not desire in these matters to engage in disputes with anyone, even on points that are disputable. My goal is this alone; that if, among errors that may abound in these con-

---

[4] *De Revolutionibus* (Nuremberg, 1543), f. iiii.

siderations of a subject remote from my profession, there is anything that may be serviceable to the holy Church in making a decision concerning the Copernican system, it may be taken and utilized as seems best to the superiors. And if not, let my book be torn and burnt, as I neither intend nor pretend to gain from it any fruit that is not pious and Catholic. And though many of the things I shall reprove have been heard by my own ears, I shall freely grant to those who have spoken them that they never said them, if that is what they wish, and I shall confess myself to have been mistaken. Hence let whatever I reply be addressed not to them, but to whoever may have held such opinions.

The reason produced for condemning the opinion that the earth moves and the sun stands still is that in many places in the Bible one may read that the sun moves and the earth stands still. Since the Bible cannot err, it follows as a necessary consequence that anyone takes an erroneous and heretical position who maintains that the sun is inherently motionless and the earth movable.

With regard to this argument, I think in the first place that it is very pious to say and prudent to affirm that the holy Bible can never speak untruth—whenever its true meaning is understood. But I believe nobody will deny that it is often very abstruse, and may say things which are quite different from what its bare words signify. Hence in expounding the Bible if one were always to confine oneself to the unadorned grammatical meaning, one might fall into error. Not only contradictions and propositions far from true might thus be made to appear in the Bible, but even grave heresies and follies. Thus it would be necessary to assign to God feet, hands, and eyes, as well as corporeal and human affections, such as anger, repentance, hatred, and sometimes even the forgetting of things past and ignorance of those to come. These propositions uttered by the Holy Ghost were set down in that manner by the sacred scribes in order to accommodate them to the capacities of the common people, who are rude and unlearned. For the sake of those who deserve to be separated from the herd, it is necessary that wise expositors should produce the true senses of such passages, together with the special reasons for which they were set down in these words. This doctrine is so widespread and so definite with all theologians that it would be superfluous to adduce evidence for it.

Hence I think that I may reasonably conclude that whenever the Bible has occasion to speak of any physical conclusion (especially those which are very abstruse and hard to understand), the rule has been observed of avoiding confusion in the minds of the common people which would render them contumacious toward the higher mysteries. Now the Bible, merely to condescend to popular capacity, has not hesitated to obscure some very important pronouncements, attributing to God himself some qualities extremely remote from (and even contrary to) His essence. Who, then, would positively declare that this principle has been set aside, and the Bible has confined itself rigorously to the bare and restricted sense of its words, when speaking but casually of the earth, of water, of the sun, or of any other created thing? Especially in view of the fact that these things in no way concern the primary purpose of the sacred writings, which is the service of God and the salvation of souls—matters infinitely beyond the comprehension of the common people.

This being granted, I think that in discussions of physical problems we ought to begin not from the authority of scriptural passages, but from sense-experiences and necessary demonstrations; for the holy Bible and the phenomena of nature proceed alike from the divine Word, the former as the dictate of the Holy Ghost and the latter as the observant executrix of God's commands. It is necessary for the Bible, in order to be accommodated to the understanding of every man, to speak many things which appear to differ from the absolute truth so far as the bare meaning of the words is concerned. But Nature, on the other hand, is inexorable and immutable; she never transgresses the laws imposed upon her, or cares a whit whether her abstruse reasons and methods of operation are understandable to men. For that reason it appears that nothing physical which sense-experience sets before our eyes, or which necessary demonstrations prove to us, ought to be called in question (much less condemned) upon the testimony of biblical passages which may have some different meaning beneath their words. For the Bible is not chained in every expression to conditions as strict as those which govern all physical effects; nor is God any less excellently revealed in Nature's actions than in the sacred statements of the Bible. Perhaps this is what Tertullian meant by these words:

"We conclude that God is known first through Nature, and then again, more particularly, by doctrine; by Nature in His works, and by doctrine in His revealed word."[5]

From this I do not mean to infer that we need not have an extraordinary esteem for the passages of holy Scripture. On the contrary, having arrived at any certainties in physics, we ought to utilize these as the most appropriate aids in the true exposition of the Bible and in the investigation of those meanings which are necessarily contained therein, for these must be concordant with demonstrated truths. I should judge that the authority of the Bible was designed to persuade men of those articles and propositions which, surpassing all human reasoning, could not be made credible by science, or by any other means than through the very mouth of the Holy Spirit.

Yet even in those propositions which are not matters of faith, this authority ought to be preferred over that of all human writings which are supported only by bare assertions or probable arguments, and not set forth in a demonstrative way. This I hold to be necessary and proper to the same extent that divine wisdom surpasses all human judgment and conjecture.

But I do not feel obliged to believe that that same God who has endowed us with senses, reason, and intellect has intended to forgo their use and by some other means to give us knowledge which we can attain by them. He would not require us to deny sense and reason in physical matters which are set before our eyes and minds by direct experi-

[5] *Adversus Marcionem,* ii, 18.

ence or necessary demonstrations. This must be especially true in those sciences of which but the faintest trace (and that consisting of conclusions) is to be found in the Bible. Of astronomy, for instance, so little is found that none of the planets except Venus are so much as mentioned, and this only once or twice under the name of "Lucifer." If the sacred scribes had had any intention of teaching people certain arrangements and motions of the heavenly bodies, or had they wished us to derive such knowledge from the Bible, then in my opinion they would not have spoken of these matters so sparingly in comparison with the infinite number of admirable conclusions which are demonstrated in that science. Far from pretending to teach us the constitution and motions of the heavens and the stars, with their shapes, magnitudes, and distances, the authors of the Bible intentionally forbore to speak of these things, though all were quite well known to them. Such is the opinion of the holiest and most learned Fathers, and in St. Augustine we find the following words:

"It is likewise commonly asked what we may believe about the form and shape of the heavens according to the Scriptures, for many contend much about these matters. But with superior prudence our authors have forborne to speak of this, as in no way furthering the student with respect to a blessed life—and, more important still, as taking up much of that time which should be spent in holy exercises. What is it to me whether heaven, like a sphere, surrounds the earth on all sides as a mass balanced in the center of the universe, or whether like a dish it merely covers and overcasts the earth? Belief in Scripture is urged rather for the reason we have often mentioned; that is, in order that no one, through ignorance of divine passages, finding anything in our Bibles or hearing anything cited from them of such a nature as may seem to oppose manifest conclusions, should be induced to suspect their truth when they teach, relate, and deliver more profitable matters. Hence let it be said briefly, touching the form of heaven, that our authors knew the truth but the Holy Spirit did not desire that men

should learn things that are useful to no one for salvation."[6]

The same disregard of these sacred authors toward beliefs about the phenomena of the celestial bodies is repeated to us by St. Augustine in his next chapter. On the question whether we are to believe that the heaven moves or stands still, he writes thus:

"Some of the brethren raise a question concerning the motion of heaven, whether it is fixed or moved. If it is moved, they say, how is it a firmament? If it stands still, how do these stars which are held fixed in it go round from east to west, the more northerly performing shorter circuits near the pole, so that heaven (if there is another pole unknown to us) may seem to revolve upon some axis, or (if there is no other pole) may be thought to move as a discus? To these men I reply that it would require many subtle and profound reasonings to find out which of these things is actually so; but to undertake this and discuss it is consistent neither with my leisure nor with the duty of those whom I desire to instruct in essential matters more directly conducing to their salvation and to the benefit of the holy Church."[7]

From these things it follows as a necessary consequence that, since the Holy Ghost did not intend to teach us whether heaven moves or stands still, whether its shape is spherical or like a discus or extended in a plane, nor whether the earth is located at its center or off to one side, then so much the less was it intended to settle for us any other conclusion of the same kind. And the motion or rest of the earth and the sun is so closely linked with the things just named, that without a determination of the one, neither side can be taken in the other matters. Now if the Holy Spirit has purposely neglected to teach us propositions of this sort as irrelevant to the highest goal (that is, to our salvation), how can anyone affirm that it is obligatory to take sides on them, and that one belief is required by faith, while the other side is erroneous? Can an opinion be heretical and yet

[6] *De Genesi ad literam* ii, 9. Galileo has noted also: "The same is to be read in Peter the Lombard, master of opinions."
[7] *Ibid.*, ii, 10.

have no concern with the salvation of souls? Can the Holy Ghost be asserted not to have intended teaching us something that does concern our salvation? I would say here something that was heard from an ecclesiastic of the most eminent degree: "That the intention of the Holy Ghost is to teach us how one goes to heaven, not how heaven goes."[8]

But let us again consider the degree to which necessary demonstrations and sense experiences ought to be respected in physical conclusions, and the authority they have enjoyed at the hands of holy and learned theologians. From among a hundred attestations I have selected the following:

"We must also take heed, in handling the doctrine of Moses, that we altogether avoid saying positively and confidently anything which contradicts manifest experiences and the reasoning of philosophy or the other sciences. For since every truth is in agreement with all other truth, the truth of Holy Writ cannot be contrary to the solid reasons and experiences of human knowledge."[9]

And in St. Augustine we read: "If anyone shall set the authority of Holy Writ against clear and manifest reason, he who does this knows not what he has undertaken; for he opposes to the truth not the meaning of the Bible, which is beyond his comprehension, but rather his own interpretation; not what is in the Bible, but what he has found in himself and imagines to be there."[10]

This granted, and it being true that two truths cannot contradict one another, it is the function of wise expositors to seek out the true senses of scriptural texts. These will unquestionably accord with the physical conclusions which manifest sense and necessary demonstrations have previously made certain to us. Now the Bible, as has been remarked, admits in many places expositions that are remote

---

[8] A marginal note by Galileo assigns this epigram to Cardinal Baronius (1538–1607). Baronius visited Padua with Cardinal Bellarmine in 1598, and Galileo probably met him at that time.

[9] Pererius on Genesis, near the beginning.

[10] In the seventh letter to Marcellinus.

from the signification of the words for reasons we have already given. Moreover, we are unable to affirm that all interpreters of the Bible speak by divine inspiration, for if that were so there would exist no differences between them about the sense of a given passage. Hence I should think it would be the part of prudence not to permit anyone to usurp scriptural texts and force them in some way to maintain any physical conclusion to be true, when at some future time the senses and demonstrative or necessary reasons may show the contrary. Who indeed will set bounds to human ingenuity? Who will assert that everything in the universe capable of being perceived is already discovered and known? Let us rather confess quite truly that "Those truths which we know are very few in comparison with those which we do not know."

We have it from the very mouth of the Holy Ghost that God delivered up the world to disputations, *so that man cannot find out the work that God hath done from the beginning even to the end.*[11] In my opinion no one, in contradiction to that dictum, should close the road to free philosophizing about mundane and physical things, as if everything had already been discovered and revealed with certainty. Nor should it be considered rash not to be satisfied with those opinions which have become common. No one should be scorned in physical disputes for not holding to the opinions which happen to please other people best, especially concerning problems which have been debated among the greatest philosophers for thousands of years. One of these is the stability of the sun and mobility of the earth, a doctrine believed by Pythagoras and all his followers, by Heracleides of Pontus[12] (who was one of them),

[11] Ecclesiastes 3:11.

[12] Heracleides was born about 390 B.C. and is said to have attended lectures by Aristotle at Athens. He believed that the earth rotated on its axis, but not that it moved around the sun. He also discovered that Mercury and Venus revolve around the sun, and may have developed a system similar to that of Tycho.

by Philolaus the teacher of Plato,[13] and by Plato himself according to Aristotle. Plutarch writes in his *Life of Numa* that Plato, when he had grown old, said it was most absurd to believe otherwise.[14] The same doctrine was held by Aristarchus of Samos,[15] as Archimedes tells us; by Seleucus[16] the mathematician, by Nicetas[17] the philosopher (on the testimony of Cicero), and by many others. Finally this opinion has been amplified and confirmed with many observations and demonstrations by Nicholas Copernicus. And Seneca,[18] a most eminent philosopher, advises us in his book on comets that we should more diligently seek to ascertain whether it is in the sky or in the earth that the diurnal rotation resides.

Hence it would probably be wise and useful counsel if, beyond articles which concern salvation and the establish-

[13] Philolaus, an early follower of Pythagoras, flourished at Thebes toward the end of the fifth century B.C. Although a contemporary of Socrates, the teacher of Plato, he had nothing to do with Plato's instruction. According to Philolaus the earth revolved around a central fire, but not about the sun (cf. note 7, p. 34).

[14] "Plato held opinion in that age, that the earth was in another place than in the very middest, and that the centre of the world, as the most honourable place, did appertain to some other of more worthy substance than the earth." (Trans. Sir Thomas North.) This tradition is no longer accepted.

[15] Aristarchus (ca. 310–230 B.C.) was the true forerunner of Copernicus in antiquity, and not the Pythagoreans as was generally believed in Galileo's time.

[16] Seleucus, who flourished about 150 B.C., is the only ancient astronomer known to have adopted the heliocentric system of Aristarchus. After his time this gave way entirely to the system founded by his contemporary Hipparchus.

[17] Nicetas is an incorrect form given by Copernicus to the name of Hicetas of Syracuse. Of this mathematician nothing is known beyond the fact that some of the ancients credited him instead of Philolaus with the astronomy which came to be associated with the Pythagoreans in general.

[18] Seneca (ca. 3–65 A.D.) was the tutor of Nero. He devoted the seventh book of his *Quaestiones Naturales* to comets. In the second chapter of this book he raised the question of the earth's rotation, and in the final chapters he appealed for patience and further investigation into such matters.

ment of our Faith, against the stability of which there is no danger whatever that any valid and effective doctrine can ever arise, men would not aggregate further articles unnecessarily. And it would certainly be preposterous to introduce them at the request of persons who, besides not being known to speak by inspiration of divine grace, are clearly seen to lack that understanding which is necessary in order to comprehend, let alone discuss, the demonstrations by which such conclusions are supported in the subtler sciences. If I may speak my opinion freely, I should say further that it would perhaps fit in better with the decorum and majesty of the sacred writings to take measures for preventing every shallow and vulgar writer from giving to his compositions (often grounded upon foolish fancies) an air of authority by inserting in them passages from the Bible, interpreted (or rather distorted) into senses as far from the right meaning of Scripture as those authors are near to absurdity who thus ostentatiously adorn their writings. Of such abuses many examples might be produced, but for the present I shall confine myself to two which are germane to these astronomical matters. The first concerns those writings which were published against the existence of the Medicean planets recently discovered by me, in which many passages of holy Scripture were cited.[19] Now that everyone has seen these planets, I should like to know what new interpretations those same antagonists employ in expounding the Scripture and excusing their own sim-

---

[19] The principal book which had offended in this regard was the *Dianoia Astronomica* . . . of Francesco Sizzi (Venice, 1611). About the time Galileo arrived at Florence, Sizzi departed for France, where he came into association with some good mathematicians. In 1613 he wrote to a friend at Rome to express his admiration of Galileo's work on floating bodies and to deride its opponents. The letter was forwarded to Galileo. In it Sizzi had reported, though rather cryptically, upon some French observations concerning sunspots, and it was probably this which led Galileo to his knowledge of the tilt of the sun's axis (cf. note 14, p. 125). Sizzi was broken on the wheel in 1617 for writing a pamphlet against the king of France.

plicity. My other example is that of a man who has lately published, in defiance of astronomers and philosophers, the opinion that the moon does not receive its light from the sun but is brilliant by its own nature.[20] He supports this fancy (or rather thinks he does) by sundry texts of Scripture which he believes cannot be explained unless his theory is true; yet that the moon is inherently dark is surely as plain as daylight.

It is obvious that such authors, not having penetrated the true senses of Scripture, would impose upon others an obligation to subscribe to conclusions that are repugnant to manifest reason and sense, if they had any authority to do so. God forbid that this sort of abuse should gain countenance and authority, for then in a short time it would be necessary to proscribe all the contemplative sciences. People who are unable to understand perfectly both the Bible and the sciences far outnumber those who do understand. The former, glancing superficially through the Bible, would arrogate to themselves the authority to decree upon every question of physics on the strength of some word which they have misunderstood, and which was employed by the sacred authors for some different purpose. And the smaller number of understanding men could not dam up the furious torrent of such people, who would gain the majority of followers simply because it is much more pleasant to gain a reputation for wisdom without effort or study than to consume oneself tirelessly in the most laborious disciplines. Let us therefore render thanks to Almighty God, who in His beneficence protects us from this danger by depriving such persons of all authority, reposing the power of consultation, decision, and decree on such important matters in the high wisdom and benevolence of most prudent

[20] This is frequently said to refer to J. C. Lagalla's *De phaenominis in orbe lunae* . . . (Venice, 1612), a wretched book which has the sole distinction of being the first to mention the word "telescope" in print. A more probable reference, however, seems to be to the *Dialogo di Fr. Ulisse Albergotti . . . nel quale si tiene . . . la Luna esser da sé luminosa . . .* (Viterbo, 1613).

Fathers, and in the supreme authority of those who cannot fail to order matters properly under the guidance of the Holy Ghost. Hence we need not concern ourselves with the shallowness of those men whom grave and holy authors rightly reproach, and of whom in particular St. Jerome said, in reference to the Bible:

"This is ventured upon, lacerated, and taught by the garrulous old woman, the doting old man, and the prattling sophist before they have learned it. Others, led on by pride, weigh heavy words and philosophize amongst women concerning holy Scripture. Others—oh, shame!—learn from women what they teach to men, and (as if that were not enough) glibly expound to others that which they themselves do not understand. I forbear to speak of those of my own profession who, attaining a knowledge of the holy Scriptures after mundane learning, tickle the ears of the people with affected and studied expressions, and declare that everything they say is to be taken as the law of God. Not bothering to learn what the prophets and the apostles have maintained, they wrest incongruous testimonies into their own senses—as if distorting passages and twisting the Bible to their individual and contradictory whims were the genuine way of teaching, and not a corrupt one."[21]

I do not wish to place in the number of such lay writers some theologians whom I consider men of profound learning and devout behavior, and who are therefore held by me in great esteem and veneration. Yet I cannot deny that I feel some discomfort which I should like to have removed, when I hear them pretend to the power of constraining others by scriptural authority to follow in a physical dispute that opinion which they think best agrees with the Bible, and then believe themselves not bound to answer the opposing reasons and experiences. In explanation and support of this opinion they say that since theology is queen of all the sciences, she need not bend in any way to accommodate herself to the teachings of less worthy sciences which are subordinate to her; these others must rather be referred to

[21] *Epistola ad Paulinum*, 103.

her as to their supreme empress, changing and altering their conclusions according to her statutes and decrees. They add further that if in the inferior sciences any conclusion should be taken as certain in virtue of demonstrations or experiences, while in the Bible another conclusion is found repugnant to this, then the professors of that science should themselves undertake to undo their proofs and discover the fallacies in their own experiences, without bothering the theologians and exegetes. For, they say, it does not become the dignity of theology to stoop to the investigation of fallacies in the subordinate sciences; it is sufficient for her merely to determine the truth of a given conclusion with absolute authority, secure in her inability to err.

Now the physical conclusions in which they say we ought to be satisfied by Scripture, without glossing or expounding it in senses different from the literal, are those concerning which the Bible always speaks in the same manner and which the holy Fathers all receive and expound in the same way. But with regard to these judgments I have had occasion to consider several things, and I shall set them forth in order that I may be corrected by those who understand more than I do in these matters—for to their decisions I submit at all times.

First, I question whether there is not some equivocation in failing to specify the virtues which entitle sacred theology to the title of "queen." It might deserve that name by reason of including everything that is learned from all the other sciences and establishing everything by better methods and with profounder learning. It is thus, for example, that the rules for measuring fields and keeping accounts are much more excellently contained in arithmetic and in the geometry of Euclid than in the practices of surveyors and accountants. Or theology might be queen because of being occupied with a subject which excels in dignity all the subjects which compose the other sciences, and because her teachings are divulged in more sublime ways.

That the title and authority of queen belongs to theology in the first sense, I think will not be affirmed by theologians who have any skill in the other sciences. None of these, I think, will say that geometry, astronomy, music, and medicine are much more excellently contained in the Bible than they are in the books of Archimedes, Ptolemy, Boethius, and Galen. Hence it seems likely that regal pre-eminence is given to theology in the second sense; that is, by reason of its subject and the miraculous communication of divine revelation of conclusions which could not be conceived by men in any other way, concerning chiefly the attainment of eternal blessedness.

Let us grant then that theology is conversant with the loftiest divine contemplation, and occupies the regal throne among sciences by dignity. But acquiring the highest authority in this way, if she does not descend to the lower and humbler speculations of the subordinate sciences and has no regard for them because they are not concerned with blessedness, then her professors should not arrogate to themselves the authority to decide on controversies in professions which they have neither studied nor practiced. Why, this would be as if an absolute despot, being neither a physician nor an architect but knowing himself free to command, should undertake to administer medicines and erect buildings according to his whim—at grave peril of his poor patients' lives, and the speedy collapse of his edifices.

Again, to command that the very professors of astronomy themselves see to the refutation of their own observations and proofs as mere fallacies and sophisms is to enjoin something that lies beyond any possibility of accomplishment. For this would amount to commanding that they must not see what they see and must not understand what they know, and that in searching they must find the opposite of what they actually encounter. Before this could be done they would have to be taught how to make one mental faculty command another, and the inferior powers the superior, so that the imagination and the will might be forced to believe the opposite of what the intellect understands. I am referring at all times to merely physical propositions, and not to supernatural things which are matters of faith.

I entreat those wise and prudent Fathers to consider with

great care the difference that exists between doctrines subject to proof and those subject to opinion. Considering the force exerted by logical deductions, they may ascertain that it is not in the power of the professors of demonstrative sciences to change their opinions at will and apply themselves first to one side and then to the other. There is a great difference between commanding a mathematician or a philosopher and influencing a lawyer or a merchant, for demonstrated conclusions about things in nature or in the heavens cannot be changed with the same facility as opinions about what is or is not lawful in a contract, bargain, or bill of exchange. This difference was well understood by the learned and holy Fathers, as proven by their having taken great pains in refuting philosophical fallacies. This may be found expressly in some of them; in particular, we find the following words of St. Augustine: "It is to be held as an unquestionable truth that whatever the sages of this world have demonstrated concerning physical matters is in no way contrary to our Bibles; hence whatever the sages teach in their books that is contrary to the holy Scriptures may be concluded without any hesitation to be quite false. And according to our ability let us make this evident, and let us keep the faith of our Lord, in whom are hidden all the treasures of wisdom, so that we neither become seduced by the verbiage of false philosophy nor frightened by the superstition of counterfeit religion."[22]

From the above words I conceive that I may deduce this doctrine: That in the books of the sages of this world there are contained some physical truths which are soundly demonstrated, and others that are merely stated; as to the former, it is the office of wise divines to show that they do not contradict the holy Scriptures. And as to the propositions which are stated but not rigorously demonstrated, anything contrary to the Bible involved by them must be held undoubtedly false and should be proved so by every possible means.

Now if truly demonstrated physical conclusions need not

[22] *De Genesi ad literam* i, 21.

be subordinated to biblical passages, but the latter must rather be shown not to interfere with the former, then before a physical proposition is condemned it must be shown to be not rigorously demonstrated—and this is to be done not by those who hold the proposition to be true, but by those who judge it to be false. This seems very reasonable and natural, for those who believe an argument to be false may much more easily find the fallacies in it than men who consider it to be true and conclusive. Indeed, in the latter case it will happen that the more the adherents of an opinion turn over their pages, examine the arguments, repeat the observations, and compare the experiences, the more they will be confirmed in that belief. And Your Highness knows what happened to the late mathematician of the University of Pisa[23] who undertook in his old age to look into the Copernican doctrine in the hope of shaking its foundations and refuting it, since he considered it false only because he had never studied it. As it fell out, no sooner had he understood its grounds, procedures, and demonstrations than he found himself persuaded, and from an opponent he became a very staunch defender of it. I might also name other mathematicians[24] who, moved by my latest discoveries, have confessed it necessary to alter the previously accepted system of the world, as this is simply unable to subsist any longer.

If in order to banish the opinion in question from the world it were sufficient to stop the mouth of a single man—as perhaps those men persuade themselves who, measuring the minds of others by their own, think it impossible that this doctrine should be able to continue to find adherents—then that would be very easily done. But things stand otherwise. To carry out such a decision it would be necessary not only to prohibit the book of Copernicus and the writings of other authors who follow the same opinion, but to ban the whole science of astronomy. Furthermore, it would be necessary to forbid men to look at the heavens, in order that

[23] Antonio Santucci (d. 1613).
[24] A marginal note by Galileo here mentions Father Clavius; cf. p. 153.

188

they might not see Mars and Venus sometimes quite near the earth and sometimes very distant, the variation being so great that Venus is forty times and Mars sixty times as large at one time as another. And it would be necessary to prevent Venus being seen round at one time and forked at another, with very thin horns; as well as many other sensory observations which can never be reconciled with the Ptolemaic system in any way, but are very strong arguments for the Copernican. And to ban Copernicus now that his doctrine is daily reinforced by many new observations and by the learned applying themselves to the reading of his book, after this opinion has been allowed and tolerated for those many years during which it was less followed and less confirmed, would seem in my judgment to be a contravention of truth, and an attempt to hide and supress her the more as she revealed herself the more clearly and plainly. Not to abolish and censure his whole book, but only to condemn as erroneous this particular proposition, would (if I am not mistaken) be a still greater detriment to the minds of men, since it would afford them occasion to see a proposition proved that it was heresy to believe. And to prohibit the whole science would be but to censure a hundred passages of holy Scripture which teach us that the glory and greatness of Almighty God are marvelously discerned in all his works and divinely read in the open book of heaven. For let no one believe that reading the lofty concepts written in that book leads to nothing further than the mere seeing of the splendor of the sun and the stars and their rising and setting, which is as far as the eyes of brutes and of the vulgar can penetrate. Within its pages are couched mysteries so profound and concepts so sublime that the vigils, labors, and studies of hundreds upon hundreds of the most acute minds have still not pierced them, even after continual investigations for thousands of years. The eyes of an idiot perceive little by beholding the external appearance of a human body, as compared with the wonderful contrivances which a careful and practiced anatomist or philosopher discovers in that same body when he seeks out the use of all those muscles, tendons, nerves, and bones; or when examining the functions of the heart and the other principal organs, he seeks the seat of the vital faculties, notes and observes the admirable structure of the sense organs, and (without ever ceasing in his amazement and delight) contemplates the receptacles of the imagination, the memory, and the understanding. Likewise, that which presents itself to mere sight is as nothing in comparison with the high marvels that the ingenuity of learned men discovers in the heavens by long and accurate observation. And that concludes what I have to say on this matter.

Next let us answer those who assert that those physical propositions of which the Bible speaks always in one way, and which the Fathers all harmoniously accept in the same sense, must be taken according to the literal sense of the words without glosses or interpretations, and held as most certain and true. The motion of the sun and stability of the earth, they say, is of this sort; hence it is a matter of faith to believe in them, and the contrary view is erroneous.

To this I wish first to remark that among physical propositions there are some with regard to which all human science and reason cannot supply more than a plausible opinion and a probable conjecture in place of a sure and demonstrated knowledge; for example, whether the stars are animate. Then there are other propositions of which we have (or may confidently expect) positive assurances through experiments, long observation, and rigorous demonstration; for example, whether or not the earth and the heavens move, and whether or not the heavens are spherical. As to the first sort of propositions, I have no doubt that where human reasoning cannot reach—and where consequently we can have no science but only opinion and faith—it is necessary in piety to comply absolutely with the strict sense of Scripture. But as to the other kind, I should think, as said before, that first we are to make certain of the fact, which will reveal to us the true senses of the Bible, and these will most certainly be found to agree with the proved fact (even though at first the words sounded otherwise), for two truths can never contradict each other. I take this to be an orthodox and indisputable doctrine, and I find it specifically in St. Augustine when he speaks of the shape of heaven and what we may believe concerning that. Astronomers seem to declare what is contrary to Scripture, for they hold the heavens to be spherical, while the Scripture calls it "stretched out like a curtain."[25] St. Augustine opines that we are not to be concerned lest the Bible contradict astronomers; we are to believe its authority if what they say is false and is founded only on the conjectures of frail humanity. But if what they say is proved by unquestionable arguments, this holy Father does not say that the astronomers are to be ordered to dissolve their proofs and declare their own conclusions to be false. Rather, he says it must be demonstrated that what is meant in the Bible by "curtain" is not contrary to their proofs. Here are his words:

"But some raise the following objection. 'How is it that the passage in our Bibles, *Who stretcheth out the heavens as a curtain,* does not contradict those who maintain the heavens to have a spherical shape?' It does contradict them if what they affirm is false, for that is true which is spoken by divine authority rather than that which proceeds from human frailty. But if, peradventure, they should be able to prove their position by experiences which place it beyond question, then it is to be demonstrated that our speaking of a curtain in no way contradicts their manifest reasons."[26]

He then proceeds to admonish us that we must be no less careful and observant in reconciling a passage of the Bible with any demonstrated physical proposition than with some other biblical passage which might appear contrary to the first. The circumspection of this saint indeed deserves admiration and imitation, when even in obscure conclusions (of which we surely can have no knowledge through human proofs) he shows great reserve in determining what is to be believed. We see this from what he writes at the end of the second book of his commentary on Genesis, concerning the question whether the stars are to be believed animate:

---

[25] Psalms 103:2 (Douay); 104:2 (King James).
[26] *De Genesi ad literam* [ii,] 9.

"Although at present this matter cannot be settled, yet I suppose that in our further dealing with the Bible we may meet with other relevant passages, and then we may be permitted, if not to determine anything finally, at least to gain some hint concerning this matter according to the dictates of sacred authority. Now keeping always our respect for moderation in grave piety, we ought not to believe anything inadvisedly on a dubious point, lest in favor of our error we conceive a prejudice against something that truth hereafter may reveal to be not contrary in any way to the sacred books of either the Old or the New Testament."

From this and other passages the intention of the holy Fathers appears to be (if I am not mistaken) that in questions of nature which are not matters of faith it is first to be considered whether anything is demonstrated beyond doubt or known by sense-experience, or whether such knowledge or proof is possible; if it is, then, being the gift of God, it ought to be applied to find out the true senses of holy Scripture in those passages which superficially might seem to declare differently. These senses would unquestionably be discovered by wise theologians, together with the reasons for which the Holy Ghost sometimes wished to veil itself under words of different meaning, whether for our exercise, or for some purpose unknown to me.

As to the other point, if we consider the primary aim of the Bible, I do not think that its having always spoken in the same sense need disturb this rule. If the Bible, accommodating itself to the capacity of the common people, has on one occasion expressed a proposition in words of different sense from the essence of that proposition, then why might it not have done the same, and for the same reason, whenever the same thing happened to be spoken of? Nay, to me it seems that not to have done this would but have increased confusion and diminished belief among the people.

Regarding the state of rest or motion of the sun and earth, experience plainly proves that in order to accommodate the common people it was necessary to assert of these things precisely what the words of the Bible convey. Even in our own age, people far less primitive continue to maintain the same opinion for reasons which will be found extremely trivial if well weighed and examined, and upon the basis of experiences that are wholly false or altogether beside the point. Nor is it worth while to try to change their opinion, they being unable to understand the arguments on the opposite side, for these depend upon observations too precise and demonstrations too subtle, grounded on abstractions which require too strong an imagination to be comprehended by them. Hence even if the stability of heaven and the motion of the earth should be more than certain in the minds of the wise, it would still be necessary to assert the contrary for the preservation of belief among the all-too-numerous vulgar. Among a thousand ordinary men who might be questioned concerning these things, probably not a single one will be found to answer anything except that it looks to him as if the sun moves and the earth stands still, and therefore he believes this to be certain. But one need not on that account take the common popular assent as an argument for the truth of what is stated; for if we should examine these very men concerning their reasons for what they believe, and on the other hand listen to the experiences and proofs which induce a few others to believe the contrary, we should find the latter to be persuaded by very sound arguments, and the former by simple appearances and vain or ridiculous impressions.

It is sufficiently obvious that to attribute motion to the sun and rest to the earth was therefore necessary lest the shallow minds of the common people should become confused, obstinate, and contumacious in yielding assent to the principal articles that are absolutely matters of faith. And if this was necessary, there is no wonder at all that it was carried out with great prudence in the holy Bible. I shall say further that not only respect for the incapacity of the vulgar, but also current opinion in those times, made the sacred authors accommodate themselves (in matters unnecessary to salvation) more to accepted usage than to the true essence of things. Speaking of this, St. Jerome writes:

"As if many things were not spoken in the Holy Bible according to the judgment of those times in which they were acted, rather than according to the truth contained."[27] And elsewhere the same saint says: "It is the custom for the biblical scribes to deliver their judgments in many things according to the commonly received opinion of their times."[28] And on the words in the twenty-sixth chapter of Job, *He stretcheth out the north over the void, and hangeth the earth above nothing*,[29] St. Thomas Aquinas notes that the Bible calls "void" or "nothing" that space which we know to be not empty, but filled with air. Nevertheless the Bible, he says, in order to accommodate itself to the beliefs of the common people (who think there is nothing in that space), calls it "void" and "nothing." Here are the words of St. Thomas: "What appears to us in the upper hemisphere of the heavens to be empty, and not a space filled with air, the common people regard as void; and it is usually spoken of in the holy Bible according to the ideas of the common people."[30]

Now from this passage I think one may very logically argue that for the same reason the Bible had still more cause to call the sun movable and the earth immovable. For if we were to test the capacity of the common people, we should find them even less apt to be persuaded of the stability of the sun and the motion of the earth than to believe that the space which environs the earth is filled with air. And if on this point it would not have been difficult to convince the common people, and yet the holy scribes forbore to attempt it, then it certainly must appear reasonable that in other and more abstruse propositions they have followed the same policy.

Copernicus himself knew the power over our ideas that is exerted by custom and by our inveterate way of conceiving things since infancy. Hence, in order not to increase for us the confusion and difficulty of abstraction, after he had

[27] On Jeremiah, ch. 28.
[28] On Matthew, ch. 13.
[29] Job 26:7.
[30] Aquinas on Job.

first demonstrated that the motions which appear to us to belong to the sun or to the firmament are really not there but in the earth, he went on calling them motions of the sun and of the heavens when he later constructed his tables to apply them to use. He thus speaks of "sunrise" and "sunset," of the "rising and setting" of the stars, of changes in the obliquity of the ecliptic and of variations in the equinoctial points, of the mean motion and variations in motion of the sun, and so on. All these things really relate to the earth, but since we are fixed to the earth and consequently share in its every motion, we cannot discover them in the earth directly, and are obliged to refer them to the heavenly bodies in which they make their appearance to us. Hence we name them as if they took place where they appear to us to take place; and from this one may see how natural it is to accommodate things to our customary way of seeing them.

Next we come to the proposition that agreement on the part of the Fathers, when they all accept a physical proposition from the Bible in the same sense, must give that sense authority to such a degree that belief in it becomes a matter of faith. I think this should be granted at most only of those propositions which have actually been discussed by the Fathers with great diligence, and debated on both sides, with them all finally concurring in the censure of one side and the adoption of the other. But the motion of the earth and stability of the sun is not an opinion of that kind, inasmuch as it was completely hidden in those times and was far removed from the questions of the schools; it was not even considered, much less adhered to, by anyone. Hence we may believe that it never so much as entered the thoughts of the Fathers to debate this. Bible texts, their own opinions, and the agreement of all men concurred in one belief, without meeting contradiction from anyone. Hence it is not sufficient to say that because all the Fathers admitted the stability of the earth, this is a matter of faith; one would have to prove also that they had condemned the contrary opinion. And I may go on to say that they left this out because they had no occasion to reflect upon the matter and discuss it; their opinion was admitted only as current, and not as analyzed and determined. I think I have very good reason for saying this.

Either the Fathers reflected upon this conclusion as controversial, or they did not; if not, then they cannot have decided anything about it even in their own minds, and their incognizance of it does not oblige us to accept teaching which they never imposed, even in intention. But if they had reflected upon and considered it, and if they judged it to be erroneous, then they would long ago have condemned it; and this they are not found to have done. Indeed, some theologians have but now begun to consider it, and they are not seen to deem it erroneous. Thus in the *Commentaries on Job* of Didacus à Stunica, where the author comments upon the words *Who moveth the earth from its place . . . ,*[31] he discourses at length upon the Copernican opinion and concludes that the mobility of the earth is not contrary to Scripture.

Besides, I question the truth of the statement that the church commands us to hold as matters of faith all physical conclusions bearing the stamp of harmonious interpretation by all the Fathers. I think this may be an arbitrary simplification of various council decrees by certain people to favor their own opinion. So far as I can find, all that is really prohibited is the "perverting into senses contrary to that of the holy Church or that of the concurrent agreement of the Fathers those passages, and those alone, which pertain to faith or ethics, or which concern the edification of Christian doctrine." So said the Council of Trent in its fourth session. But the mobility or stability of the earth or sun is neither a matter of faith nor one contrary to ethics. Neither would anyone pervert passages of Scripture in opposition to the holy Church or to the Fathers, for those who have written on this matter have never employed scriptural passages. Hence it remains the office of grave and wise theologians to interpret the passages according to their true meaning.

[31] Job 9:6. The commentary was that of Didacus à Stunica, published at Toledo in 1584; cf. p. 219.

Council decrees are indeed in agreement with the holy Fathers in these matters, as may be seen from the fact that they abstain from enjoining us to receive physical conclusions as matters of faith, and from censuring the opposite opinions as erroneous. Attending to the primary and original intention of the holy Church, they judge it useless to be occupied in attempting to get to the bottom of such matters. Let me remind Your Highness again of St. Augustine's reply to those brethren who raised the question whether the heavens really move or stand still: "To these men I reply that it would require many subtle and profound reasonings to find out which of these things is actually so; but to undertake this and discuss it is consistent neither with my leisure nor with the duty of those whom I desire to instruct in essential matters more directly conducive to their salvation and to the benefit of the holy Church."[32]

Yet even if we resolved to condemn or admit physical propositions according to scriptural passages uniformly expounded in the same sense by all the Fathers, I still fail to see how that rule can apply in the present case, inasmuch as diverse expositions of the same passage occur among the Fathers. Dionysius the Areopagite says that it is the *primum mobile*[33] which stood still, not the sun.[34] St. Augustine is of the same opinion; that is, that all celestial bodies would be stopped; and the Bishop of Avila concurs.[35] What is

[32] Cf. note 6, p. 185.

[33] The outermost crystalline sphere was known as the *primum mobile,* or prime mover, and was supposed to complete each revolution in twenty-four hours, causing night and day. A part of its motion was imagined to be transmitted to each inner sphere, sweeping along the fixed stars and the planets (which included the sun and moon) at nearly its own speed. The inherent motion of the other spheres was supposed to be eastward at much slower rates. In the case of the sun, this speed would have the same proportion to that of the *primum mobile* as a day has to a year.

[34] In the *Epistola ad Polycarpum.*

[35] In the second book of St. Augustine's *De Mirabilius Sacrae Scripturae.* The Bishop of Avila referred to was Alfonso

197

more, among the Jewish authors endorsed by Josephus,[36] some held that the sun did not really stand still, but that it merely appeared to do so by reason of the shortness of the time during which the Israelites administered defeat to their enemies. (Similarly, with regard to the miracle in the time of Hezekiah, Paul of Burgos was of the opinion that this took place not in the sun but on the sundial.)[37] And as a matter of fact no matter what system of the universe we assume, it is still necessary to gloss and interpret the words in the text of Joshua, as I shall presently show.

But finally let us grant to these gentlemen even more than they demand; namely, let us admit that we must subscribe entirely to the opinion of wise theologians. Then, since this particular dispute does not occur among the ancient Fathers, it must be undertaken by the wise men of this age. After first hearing the experiences, observations, arguments, and proofs of philosophers and astronomers on both sides—for the controversy is over physical problems and logical dilemmas, and admits of no third alternative —they will be able to determine the matter positively, in accordance with the dictates of divine inspiration. But as to those men who do not scruple to hazard the majesty and dignity of holy Scripture to uphold the reputation of their own vain fancies, let them not hope that a decision such as this is to be made without minutely airing and discussing all the arguments on both sides. Nor need we fear this from men who will make it their whole business to examine most attentively the very foundations of this doctrine, and who will do so only in a holy zeal for the truth, the Bible, and the majesty, dignity, and authority in which every Christian wants to see these maintained.

Anyone can see that dignity is most desired and best secured by those who submit themselves absolutely to the

---

Tostado (1400–55), and the reference is to his twenty-second and twenty-fourth questions on the tenth chapter of Joshua.

[36] Flavius Josephus (ca. 37–95 A.D.), historian of the Jews.

[37] Isaiah 38:8. Paul of Burgos (ca. 1350–1435), also known as Paul de Santa Maria, was a Jewish convert to Christianity who became Bishop of Burgos.

holy Church and do not demand that one opinion or another be prohibited, but merely ask the right to propose things for consideration which may the better guarantee the soundest decision—not by those who, driven by personal interest or stimulated by malicious hints, preach that the Church should flash her sword without delay simply because she has the power to do so. Such men fail to realize that it is not always profitable to do everything that lies within one's power. The most holy Fathers did not share their views. They knew how prejudicial (and how contrary to the primary intention of the Catholic Church) it would be to use scriptural passages for deciding physical conclusions, when either experiments or logical proofs might in time show the contrary of what the literal sense of the words signifies. Hence they not only proceeded with great circumspection, but they left the following precepts for the guidance of others: "In points that are obscure, or far from clear, if we should read anything in the Bible that may allow of several constructions consistently with the faith to be taught, let us not commit ourselves to any one of these with such precipitous obstinacy that when, perhaps, the truth is more diligently searched into, this may fall to the ground, and we with it. Then we would indeed be seen to have contended not for the sense of divine Scripture, but for our own ideas by wanting something of ours to be the sense of Scripture when we should rather want the meaning of Scripture to be ours."[38] And later it is added, to teach us that no proposition can be contrary to the faith unless it has first been proven to be false: "A thing is not forever contrary to the faith until disproved by most certain truth. When that happens, it was not holy Scripture that ever affirmed it, but human ignorance that imagined it."

From this it is seen that the interpretation which we impose upon passages of Scripture would be false whenever it disagreed with demonstrated truths. And therefore we should seek the incontrovertible sense of the Bible with the

---

[38] This and the ensuing quotations from St. Augustine are referred to *De Genesi ad literam* i, 18 and 19.

assistance of demonstrated truth, and not in any way try to force the hand of Nature or deny experiences and rigorous proofs in accordance with the mere sound of words that may appeal to our frailty. Let Your Highness note further how circumspectly this saint proceeds before affirming any interpretation of Scripture to be certain and secure from all disturbing difficulties. Not content that some given sense of the Bible agrees with some demonstration, he adds: "But when some truth is demonstrated to be certain by reason, it is still not certain whether in these words of holy Scripture the writer intended this idea, or some other that is no less true. And if the context of his words prove that he did not intend this truth, the one that he did intend will not thereby be false, but most true, and still more profitable for us to know." Our admiration of the circumspection of this pious author only grows when he adds the following words, being not completely convinced after seeing that logical proof, the literal words of the Bible, and all the context before and after them harmonize in the same thing: "But if the context supplies nothing to disprove this to be the author's sense, it yet remains for us to inquire whether he may not intend the other as well." Nor even yet does he resolve to accept this one interpretation and reject the other, appearing never to be able to employ sufficient caution, for he continues: "But if we find that the other also may be meant, it may be inquired which of them the writer would want to have stand, or which one he probably meant to aim at, when the true circumstances on both sides are weighed." And finally he supplies a reason for this rule of his, by showing us the perils to which those men expose the Bible and the Church, who, with more regard for the support of their own errors than for the dignity of the Bible, attempt to stretch its authority beyond the bounds which it prescribes to itself. The following words which he adds should alone be sufficient to repress or moderate the excessive license which some men arrogate to themselves: "It often falls out that a Christian may not fully understand some point about the earth, the sky, or the other elements of this world—the motion, rotation, magnitude, and distances of the stars; the known vagaries of the sun and moon; the circuits of the years and epochs; the nature of animals, fruits, stones, and other things of that sort, and hence may not expound it rightly or make it clear by experiences. Now it is too absurd, yea, most pernicious and to be avoided at all costs, for an infidel to find a Christian so stupid as to argue these matters as if they were Christian doctrine; he will scarce be able to contain his laughter at seeing error written in the skies, as the proverb says. The worst of the matter is not that a person in error should be laughed at, but that our authors should be thought by outsiders to hold the same opinions, and should be censured and rejected as ignorant, to the great prejudice of those whose salvation we are seeking. For when infidels refute any Christian on a matter which they themselves thoroughly understand, they thereby evince their slight esteem for our Bible. And why should the Bible be believed concerning the resurrection of the dead, the hope of eternal life, and the Kingdom of Heaven, when it is considered to be erroneously written as to points which admit of direct demonstration or unquestionable reasoning?"

There are men who, in defense of propositions which they do not understand, apply—and in a way commit—some text of the Bible, and then proceed to magnify their original error by adducing other passages that are even less understood than the first. The extent to which truly wise and prudent Fathers are offended by such men is declared by the same saint in the following terms: "Inexpressible trouble and sorrow are brought by rash and presumptuous men upon their more prudent brethren. When those who respect the authority of our Bible commence to reprove and refute their false and unfounded opinions, such men defend what they have put forth quite falsely and rashly by citing the Bible in their own support, repeating from memory biblical passages which they arbitrarily force to their purposes, without knowing either what they mean or to what they properly apply."

It seems to me that we may number among such men those who, being either unable or unwilling to comprehend the experiences and proofs used in support of the new doctrine by its author and his followers, nevertheless expect to bring the Scriptures to bear on it. They do not consider that the more they cite these, and the more they insist that they are perfectly clear and admit of no other interpretations than those which they put on them, the more they prejudice the dignity of the Bible—or would, if their opinion counted for anything—in the event that later truth shows the contrary and thus creates confusion among those outside the holy Church. And of these she is very solicitous, like a mother desiring to recover her children into her lap.

Your Highness may thus see how irregularly those persons proceed who in physical disputes arrange scriptural passages (and often those ill-understood by them) in the front rank of their arguments. If these men really believe themselves to have the true sense of a given passage, it necessarily follows that they believe they have in hand the absolute truth of the conclusion they intend to debate. Hence they must know that they enjoy a great advantage over their opponents, whose lot it is to defend the false position; and he who maintains the truth will have many sense-experiences and rigorous proofs on his side, whereas his antagonist cannot make use of anything but illusory appearances, quibbles, and fallacies. Now if these men know they have such advantages over the enemy even when they stay within proper bounds and produce no weapons other than those proper to philosophy, why do they, in the thick of battle, betake themselves to a dreadful weapon which cannot be turned aside, and seek to vanquish the opponent by merely exhibiting it? If I may speak frankly, I believe they have themselves been vanquished, and, feeling unable to stand up against the assaults of the adversary, they seek ways of holding him off. To that end they would forbid him the use of reason, divine gift of Providence, and would abuse the just authority of holy Scripture—which, in the general opinion of theologians, can never oppose manifest experiences and necessary demonstrations when rightly understood and applied. If I am correct, it will stand them in no stead to go running to the Bible to cover up their inability to understand (let alone resolve) their opponents' arguments, for the opinion which they fight has never been condemned by the holy Church. If they wish to proceed in sincerity, they should by silence confess themselves unable to deal with such matters. Let them freely admit that although they may argue that a position is false, it is not in their power to censure a position as erroneous—or in the power of anyone except the Supreme Pontiff, or the Church Councils. Reflecting upon this, and knowing that a proposition cannot be both true and heretical, let them employ themselves in the business which is proper to them; namely, demonstrating its falsity. And when that is revealed, either there will no longer be any necessity to prohibit it (since it will have no followers), or else it may safely be prohibited without the risk of any scandal.

Therefore let these men begin to apply themselves to an examination of the arguments of Copernicus and others, leaving condemnation of the doctrine as erroneous and heretical to the proper authorities. Among the circumspect and most wise Fathers, and in the absolute wisdom of one who cannot err, they may never hope to find the rash decisions into which they allow themselves to be hurried by some particular passion or personal interest. With regard to this opinion, and others which are not directly matters of faith, certainly no one doubts that the Supreme Pontiff has always an absolute power to approve or condemn; but it is not in the power of any created being to make things true or false, for this belongs to their own nature and to the fact. Therefore in my judgment one should first be assured of the necessary and immutable truth of the fact, over which no man has power. This is wiser counsel than to condemn either side in the absence of such certainty, thus depriving oneself of continued authority and ability to choose by determining things which are now undetermined and open and still lodged in the will of supreme authority. And in brief, if it is impossible for a conclusion to be declared heretical while we remain in doubt as to its truth, then these men are wasting their time clamoring for condemnation of the motion of the earth and stability of the sun, which they have not yet demonstrated to be impossible or false.

Now let us consider the extent to which it is true that the famous passage in Joshua may be accepted without altering the literal meaning of its words, and under what conditions the day might be greatly lengthened by obedience of the sun to Joshua's command that it stand still.

If the celestial motions are taken according to the Ptolemaic system, this could never happen at all. For the movement of the sun through the ecliptic is from west to east, and hence it is opposite to the movement of the *primum mobile*, which in that system causes day and night. Therefore it is obvious that if the sun should cease its own proper motion, the day would become shorter, and not longer. The way to lengthen the day would be to speed up the sun's proper motion; and to cause the sun to remain above the horizon for some time in one place without declining towards the west, it would be necessary to hasten this motion until it was equal to that of the *primum mobile*. This would amount to accelerating the customary speed of the sun about three hundred sixty times. Therefore if Joshua had intended his words to be taken in their pure and proper sense, he would have ordered the sun to accelerate its own motion in such a way that the impulse from the *primum mobile* would not carry it westward. But since his words were to be heard by people who very likely knew nothing of any celestial motions beyond the great general movement from east to west, he stooped to their capacity and spoke according to their understanding, as he had no intention of teaching them the arrangement of the spheres, but merely of having them perceive the greatness of the miracle. Possibly it was this consideration that first moved Dionysius the Areopagite to say that in this miracle it was the *primum mobile* that stood still, and that when this halted, all the celestial spheres stopped as a consequence —an opinion held by St. Augustine himself, and confirmed in detail by the Bishop of Avila. And indeed Joshua did intend the whole system of celestial spheres to stand still, as may be deduced from his simultaneous command to the moon, which had nothing to do with lengthening the day. And under his command to the moon we are to understand the other planets as well, though they are passed over in silence here as elsewhere in the Bible, which was not written to teach us astronomy.

It therefore seems very clear to me that if we were to accept the Ptolemaic system it would be necessary to interpret the words in some sense different from their strict meaning. Admonished by the useful precepts of St. Augustine, I shall not affirm this to be necessarily the above sense, as someone else may think of another that is more proper and harmonious. But I wish to consider next whether this very event may not be understood more consistently with what we read in the Book of Joshua in terms of the Copernican system, adding a further observation recently pointed out by me in the body of the sun. Yet I speak always with caution and reserve, and not with such great affection for my own inventions as to prefer them above those of others, or in the belief that nothing can be brought forth that will be still more in conformity with the intention of the Bible.

Suppose, then, that in the miracle of Joshua the whole system of celestial rotations stood still, in accordance with the opinion of the authors named above. Now in order that all the arrangements should not be disturbed by stopping only a single celestial body, introducing great disorder throughout the whole of Nature, I shall next assume that the sun, though fixed in one place, nevertheless revolves upon its own axis, making a complete revolution in about a month, as I believe is conclusively proven in my *Letters on Sunspots*. With our own eyes we see this movement to be slanted toward the south in the more remote part of the sun's globe, and in the nearer part to tilt toward the north, in just the same manner as all the revolutions of the planets occur. Third, if we consider the nobility of the sun, and the fact that it is the font of light which (as I shall conclusively prove) illuminates not only the moon and the earth but all the other planets, which are inherently dark, then I believe that it will not be entirely unphilosophical to say that the sun, as the chief minister of Nature and in a certain sense the heart and soul of the universe, infuses by its own rotation not only light but also motion into other bodies which surround it. And just as if the motion of the heart should cease in an animal, all other motions of its members would also cease, so if the rotation of the sun were to stop, the rotations of all the planets would stop too. And though I could produce the testimonies of many grave authors to prove the admirable power and energy of the sun, I shall content myself with a single passage from the blessed Dionysius the Areopagite in his book *Of the Divine Name*,[39] who writes thus of the sun: "His light gathers and converts to himself all things which are seen, moved, lighted, or heated; and in a word all things which are preserved by his splendor. For this reason the sun is called HELIOS, because he collects and gathers all dispersed things." And shortly thereafter he says: "This sun which we see remains one, and despite the variety of essences and qualities of things which fall under our senses, he bestows his light equally on them, and renews, nourishes, defends, perfects, divides, conjoins, cherishes, makes fruitful, increases, changes, fixes, produces, moves, and fashions all living creatures. Everything in this universe partakes of one and the same sun by His will, and the causes of many things which are shared from him are equally anticipated in him. And for so much the more reason," and so on.

The sun, then, being the font of light and the source of motion, when God willed that at Joshua's command the whole system of the world should rest and should remain for many hours in the same state, it sufficed to make the sun stand still. Upon its stopping all the other revolutions ceased; the earth, the moon, and the sun remained in the same arrangement as before, as did all the planets; nor in all that time did day decline towards night, for day was miraculously prolonged. And in this manner, by the stopping of the sun, without altering or in the least disturbing the other aspects and mutual positions of the stars, the day

[39] The book *Of the Divine Name*, then attributed to Dionysius the disciple of Paul, actually belongs to the late fifth or early sixth century.

could be lengthened on earth—which agrees exquisitely with the literal sense of the sacred text.

But if I am not mistaken, something of which we are to take no small account is that by the aid of this Copernican system we have the literal, open, and easy sense of another statement that we read in this same miracle, that the sun stood still *in the midst of the heavens*.[40] Grave theologians raise a question about this passage, for it seems very likely that when Joshua requested the lengthening of the day, the sun was near setting and not at the meridian. If the sun had been at the meridian, it seems improbable that it was necessary to pray for a lengthened day in order to pursue victory in battle, the miracle having occurred around the summer solstice when the days are longest, and the space of seven hours remaining before nightfall being sufficient. Thus grave divines have actually held that the sun was near setting, and indeed the words themselves seem to say so: *Sun, stand thou still, stand thou still*.[41] For if it had been near the meridian, either it would have been needless to request a miracle, or it would have been sufficient merely to have prayed for some retardation. Cajetan[42] is of this opinion, to which Magellan[43] subscribes, confirming it with the remark that Joshua had already done too many things that day before commanding the sun to stand still for him to have done them in half a day. Hence they are forced to interpret the words *in the midst of the heavens* a little knottily, saying that this means no more than that the sun stood still while it was in our hemisphere; that is, above our horizon. But unless I am mistaken we may avoid this and all other knots if, in agreement with the Copernican system, we place the sun in the "midst"—that is, in the center—of the celestial orbs and planetary rotations, as it is most necessary to do. Then take

[40] Joshua 10:13.
[41] Joshua 10:12.
[42] Thomas de Vio (1468–1534), Bishop of Gaeta, commenting on the *Summa Theologica* of Thomas Aquinas.
[43] Cosme Magalhaens (1553–1624), a Portuguese Jesuit who in 1612 had published a two-volume treatise on the Book of Joshua.

any hour of the day, either noon, or any hour as close to evening as you please, and the day would be lengthened and all the celestial revolutions stopped by the sun's standing still *in the midst of the heavens;* that is, in the center, where it resides. This sense is much better accommodated to the words, quite apart from what has already been said; for if the desired statement was that the sun was stopped at midday, the proper expression would have been that it "stood still at noonday," or "in the meridian circle," and not "in the midst of the heavens." For the true and only "midst" of a spherical body such as the sky is its center.

As to other scriptural passages which seem to be contrary to this opinion, I have no doubt that if the opinion itself were known to be true and proven, those very theologians who, so long as they deem it false, hold these passages to be incapable of harmonious exposition with it, would find interpretations for them which would agree very well, and especially if they would add some knowledge of astronomical science to their knowledge of divinity. At present, while they consider it false, they think they find in Scripture only passages that contradict it; but if they once entertained a different view of the matter they would probably find as many more that would harmonize with it. And then they might judge that it is fitting for the holy Church to tell that God placed the sun in the center of heaven, and that by rotating it like a wheel gave to the moon and the other wandering stars their appointed courses, when she sings the hymn:

> Most Holy God of Heaven
> Who paints with fiery splendor
> The brilliant center of the pole
> Enriched with beauteous light;
> Who, creating on the fourth day
> The flaming disk of the sun
> Gave order to the moon
> And wandering courses to the stars . . .[44]

[44] From the hymn *God, Creator of All,* attributed to St. Ambrose.

And they could say that the name "firmament" agrees literally quite well with the starry sphere and all that lies beyond the revolutions of the planets, which according to this arrangement is quite firm and immovable. Again, with the earth turning, they might think of its poles when they read *He had not yet made the earth, the rivers, and the hinges of the terrestrial orb*,[45] for hinges would seem to be ascribed in vain to the earth unless it needed them to turn upon.

[45] Proverbs 8:26 (Douay). At present the word in question is translated "poles."

5

Galileo Galilei, DIALOGUES CONCERNING TWO NEW SCIENCES
                (Trans. by Henry Crew & Alfonso de Salvio)

Isaac Newton,   PRINCIPIA MATHEMATICAS

1. Write the simple mathematical formula relating dis-
    tance, speed, and time, which you learned by the
    seventh grade, and which contains all of the infor-
    mation in the section on uniform motion. Why do
    you suppose that Galileo did not make use of such a
    concise representation?

2. In paragraph 206, Salviati describes an experiment
    as it happens and infers what would happen if no
    frictional losses were present. Such arguments are
    commonplace in physics and are essential to any
    physical analysis. Why?

3. Modern historians of science have raised questions
    concerning the probability that Galileo actually
    performed the experiment involving the bronze ball
    and the inclined plane. Do you see anything re-
    markable in the description of the experiment? Can
    you write down a mathematical expression which sum-
    marizes the results of Galileo's experiment?

4. What fundamental assumption regarding the nature of
    the universe do Galileo and Newton share?

5. How does Newton describe time and space? Why do
    you suppose he found it necessary to be so careful
    regarding these definitions? How do Newton's defi-
    nitions relate to your own concepts of time and
    space?

6. What are the implications of Newton's first rule of
    reasoning in philosophy for choosing between two
    theories which purport to explain the same set of
    phenomena?

7. How would Bacon have viewed Newton's "Rules of Rea-
    soning in Philosophy"?

8. Why do you suppose that Galileo, Newton, and the
    ancients were so fascinated with the science of
    motion?

The *Dialogues Concerning Two New Sciences*, written after the church's condemnation of Copernicus' views and published in the Protestant Netherlands in 1638, summarizes Galileo's important new theories and experiments concerning motion.

THIRD DAY

[190]

CHANGE OF POSITION. [*De Motu Locali*]

Y purpose is to set forth a very new science dealing with a very ancient subject. There is, in nature, perhaps nothing older than motion, concerning which the books written by philosophers are neither few nor small; nevertheless I have discovered by experiment some properties of it which are worth knowing and which have not hitherto been either observed or demonstrated. Some superficial observations have been made, as, for instance, that the free motion [*naturalem motum*] of a heavy falling body is continuously accelerated; * but to just what extent this acceleration occurs has not yet been announced; for so far as I know, no one has yet pointed out that the distances traversed, during equal intervals of time, by a body falling from rest, stand to one another in the same ratio as the odd numbers beginning with unity.†

It has been observed that missiles and projectiles describe a curved path of some sort; however no one has pointed out the fact that this path is a parabola. But this and other facts, not few in number or less worth knowing, I have succeeded in proving; and what I consider more important, there have been opened up to this vast and most excellent science, of which my

---

* "Natural motion" of the author has here been translated into "free motion"—since this is the term used to-day to distinguish the "natural" from the "violent" motions of the Renaissance. [*Trans.*]

† A theorem demonstrated on p. 175 below. [*Trans.*]

work is merely the beginning, ways and means by which other minds more acute than mine will explore its remote corners.

This discussion is divided into three parts; the first part deals with motion which is steady or uniform; the second treats of motion as we find it accelerated in nature; the third deals with the so-called violent motions and with projectiles.

[191]
UNIFORM MOTION

In dealing with steady or uniform motion, we need a single definition which I give as follows:

DEFINITION

By steady or uniform motion, I mean one in which the distances traversed by the moving particle during any equal intervals of time, are themselves equal.

CAUTION

We must add to the old definition (which defined steady motion simply as one in which equal distances are traversed in equal times) the word "any," meaning by this, all equal intervals of time; for it may happen that the moving body will traverse equal distances during some equal intervals of time and yet the distances traversed during some small portion of these time-intervals may not be equal, even though the time-intervals be equal.

From the above definition, four axioms follow, namely:

AXIOM I

In the case of one and the same uniform motion, the distance traversed during a longer interval of time is greater than the distance traversed during a shorter interval of time.

AXIOM II

In the case of one and the same uniform motion, the time required to traverse a greater distance is longer than the time required for a less distance.

THIRD DAY

Axiom III

In one and the same interval of time, the distance traversed at a greater speed is larger than the distance traversed at a less speed.

Axiom IV

The speed required to traverse a longer distance is greater than that required to traverse a shorter distance during the same time-interval.

Theorem I, Proposition I

If a moving particle, carried uniformly at a constant speed, traverses two distances the time-intervals required are to each other in the ratio of these distances.

Let a particle move uniformly with constant speed through two distances AB, BC, and let the time required to traverse AB be represented by DE; the time required to traverse BC, by EF;

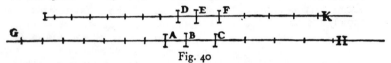

Fig. 40

then I say that the distance AB is to the distance BC as the time DE is to the time EF.

Let the distances and times be extended on both sides towards G, H and I, K; let AG be divided into any number whatever of spaces each equal to AB, and in like manner lay off in DI exactly the same number of time-intervals each equal to DE. Again lay off in CH any number whatever of distances each equal to BC; and in FK exactly the same number of time-intervals each equal to EF; then will the distance BG and the time EI be equal and arbitrary multiples of the distance BA and the time ED; and likewise the distance HB and the time KE are equal and arbitrary multiples of the distance CB and the time FE.

And since DE is the time required to traverse AB, the whole time

214

time EI will be required for the whole distance BG, and when the motion is uniform there will be in EI as many time-intervals each equal to DE as there are distances in BG each equal to BA; and likewise it follows that KE represents the time required to traverse HB.

Since, however, the motion is uniform, it follows that if the distance GB is equal to the distance BH, then must also the time IE be equal to the time EK; and if GB is greater than BH, then also IE will be greater than EK; and if less, less.* There

[193]

are then four quantities, the first AB, the second BC, the third DE, and the fourth EF; the time IE and the distance GB are arbitrary multiples of the first and the third, namely of the distance AB and the time DE.

But it has been proved that *both* of these latter quantities are either equal to, greater than, or less than the time EK and the space BH, which are arbitrary multiples of the second and the fourth. Therefore the first is to the second, namely the distance AB is to the distance BC, as the third is to the fourth, namely the time DE is to the time EF.   Q. E. D.

THEOREM II, PROPOSITION II

If a moving particle traverses two distances in equal intervals of time, these distances will bear to each other the
same ratio as the speeds. And conversely if the distances
are as the speeds then the times are equal.

Referring to Fig. 40, let AB and BC represent the two distances traversed in equal time-intervals, the distance AB for instance with the velocity DE, and the distance BC with the velocity EF. Then, I say, the distance AB is to the distance BC as the velocity DE is to the velocity EF. For if equal multiples of both distances and speeds be taken, as above, namely, GB and IE of AB and DE respectively, and in like manner HB and KE of BC and EF, then one may infer, in the same manner as above, that the multiples GB and IE are either less than, equal

---

* The method here employed by Galileo is that of Euclid as set forth in the famous 5th Definition of the Fifth Book of his *Elements*, for which see *art. Geometry* Ency. Brit. 11th Ed. p. 683. [*Trans.*]

to, or greater than equal multiples of BH and EK. Hence the theorem is established.

THEOREM III, PROPOSITION III

In the case of unequal speeds, the time-intervals required to traverse a given space are to each other inversely as the speeds.

Let the larger of the two unequal speeds be indicated by A; the smaller, by B; and let the motion corresponding to both traverse the given space CD. Then I say the time required to traverse the distance CD at speed A is to the time required to traverse the same distance at speed B, as the speed B is to the speed A. For let CD be to CE as A is to B; then, from the preceding, it follows that the time required to complete the distance CD at speed A is the same as

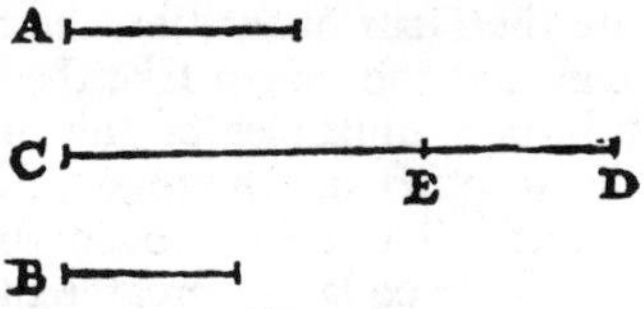

Fig. 41

[194]

the time necessary to complete CE at speed B; but the time needed to traverse the distance CE at speed B is to the time required to traverse the distance CD at the same speed as CE is to CD; therefore the time in which CD is covered at speed A is to the time in which CD is covered at speed B as CE is to CD, that is, as speed B is to speed A. Q. E. D.

THEOREM IV, PROPOSITION IV

If two particles are carried with uniform motion, but each with a different speed, the distances covered by them during unequal intervals of time bear to each other the compound ratio of the speeds and time intervals.

Let the two particles which are carried with uniform motion be E and F and let the ratio of the speed of the body E be to that of the body F as A is to B; but let the ratio of the time consumed by the motion of E be to the time consumed by the motion of F as C is to D. Then, I say, that the distance covered by E, with speed A in time C, bears to the space traversed by F with speed B

216

B in time D a ratio which is the product of the ratio of the speed A to the speed B by the ratio of the time C to the time D. For if G is the distance traversed by E at speed A during the time-interval C, and if G is to I as the speed A is to the speed B; and if also the time-interval C is to the time-interval D as I is to L, then it follows that I is the distance traversed by F in the same time that G is traversed by E since G is to I in the same ratio as the speed A to the speed B. And since I is to L in the same ratio as the time-intervals C and D, if I is the distance traversed by F during the interval C, then L will be the distance traversed by F during the interval D at the speed B.

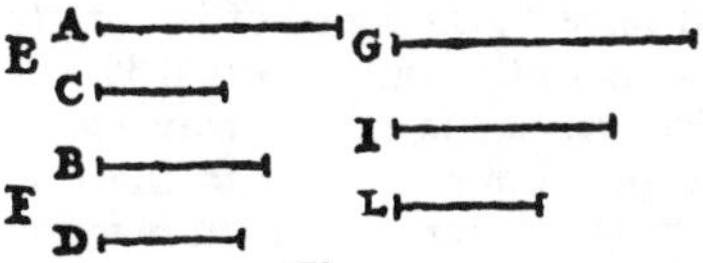

Fig. 42

But the ratio of G to L is the product of the ratios G to I and I to L, that is, of the ratios of the speed A to the speed B and of the time-interval C to the time-interval D.   Q. E. D.

[195]
THEOREM V, PROPOSITION V

If two particles are moved at a uniform rate, but with un-equal speeds, through unequal distances, then the ratio of the time-intervals occupied will be the product of the ratio of the distances by the inverse ratio of the speeds.

Let the two moving particles be denoted by A and B, and let the speed of A be to the speed of B in the ratio of V to T; in like manner let the distances trav-ersed be in the ratio

Fig. 43

of S to R; then I say that the ratio of the time-interval during which the motion of A occurs to the time-interval occupied by the motion of B is the product of the ratio of the speed T to the speed V by the ratio of the distance S to the distance R.

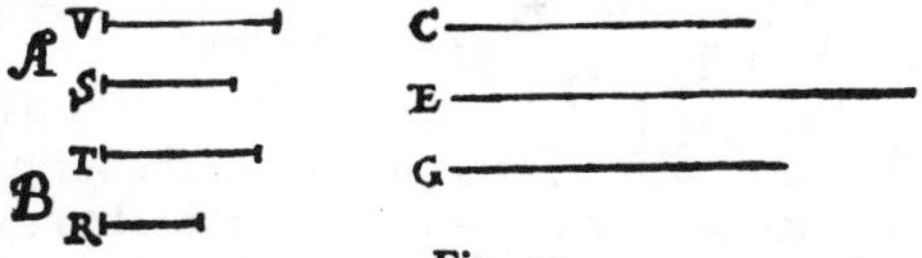

Let C be the time-interval occupied by the motion of A, and let

let the time-interval C bear to a time-interval E the same ratio as the speed T to the speed V.

And since C is the time-interval during which A, with speed V, traverses the distance S and since T, the speed of B, is to the speed V, as the time-interval C is to the time-interval E, then E will be the time required by the particle B to traverse the distance S. If now we let the time-interval E be to the time-interval G as the distance S is to the distance R, then it follows that G is the time required by B to traverse the space R. Since the ratio of C to G is the product of the ratios C to E and E to G (while also the ratio of C to E is the inverse ratio of the speeds of A and B respectively, i. e., the ratio of T to V); and since the ratio of E to G is the same as that of the distances S and R respectively, the proposition is proved.

[196]
Theorem VI, Proposition VI

If two particles are carried at a uniform rate, the ratio of their speeds will be the product of the ratio of the distances traversed by the inverse ratio of the time-intervals occupied.

Let A and B be the two particles which move at a uniform rate; and let the respective distances traversed by them have the ratio of V to T, but let the time-intervals be as S to R. Then I say the speed of A will bear to the speed of B a ratio which is the product of the ratio of the distance V to the distance T and the time-interval R to the time-interval S.

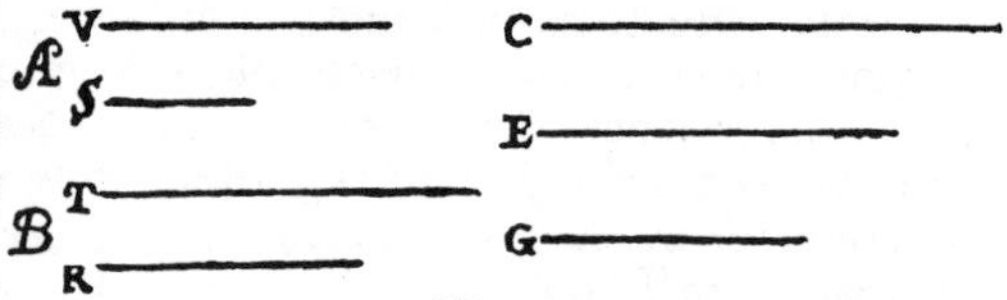

Fig. 44

Let C be the speed at which A traverses the distance V during the time-interval S; and let the speed C bear the same ratio to another speed E as V bears to T; then E will be the speed at which B traverses the distance T during the time-interval S. If now the speed E is to another speed G as the time-interval R is to the time-interval S, then G will be the speed at which the

particle

particle B traverses the distance T during the time-interval R. Thus we have the speed C at which the particle A covers the distance V during the time S and also the speed G at which the particle B traverses the distance T during the time R. The ratio of C to G is the product of the ratio C to E and E to G; the ratio of C to E is by definition the same as the ratio of the distance V to distance T; and the ratio of E to G is the same as the ratio of R to S. Hence follows the proposition.

SALV. The preceding is what our Author has written concerning uniform motion. We pass now to a new and more discriminating consideration of naturally accelerated motion, such as that generally experienced by heavy falling bodies; following is the title and introduction.

[197]
NATURALLY ACCELERATED MOTION

The properties belonging to uniform motion have been discussed in the preceding section; but accelerated motion remains to be considered.

And first of all it seems desirable to find and explain a definition best fitting natural phenomena. For anyone may invent an arbitrary type of motion and discuss its properties; thus, for instance, some have imagined helices and conchoids as described by certain motions which are not met with in nature, and have very commendably established the properties which these curves possess in virtue of their definitions; but we have decided to consider the phenomena of bodies falling with an acceleration such as actually occurs in nature and to make this definition of accelerated motion exhibit the essential features of observed accelerated motions. And this, at last, after repeated efforts we trust we have succeeded in doing. In this belief we are confirmed mainly by the consideration that experimental results are seen to agree with and exactly correspond with those properties which have been, one after another, demonstrated by us. Finally, in the investigation of naturally accelerated motion we were led, by hand as it were, in following the habit and custom of
nature

nature herself, in all her various other processes, to employ only those means which are most common, simple and easy.

For I think no one believes that swimming or flying can be accomplished in a manner simpler or easier than that instinctively employed by fishes and birds.

When, therefore, I observe a stone initially at rest falling from an elevated position and continually acquiring new increments of speed, why should I not believe that such increases take place in a manner which is exceedingly simple and rather obvious to everybody? If now we examine the matter carefully we find no addition or increment more simple than that which repeats itself always in the same manner. This we readily understand when we consider the intimate relationship between time and motion; for just as uniformity of motion is defined by and conceived through equal times and equal spaces (thus we call a motion uniform when equal distances are traversed during equal time-intervals), so also we may, in a similar manner, through equal time-intervals, conceive additions of speed as taking place without complication; thus we may picture to our

[198]

mind a motion as uniformly and continuously accelerated when, during any equal intervals of time whatever, equal increments of speed are given to it. Thus if any equal intervals of time whatever have elapsed, counting from the time at which the moving body left its position of rest and began to descend, the amount of speed acquired during the first two time-intervals will be double that acquired during the first time-interval alone; so the amount added during three of these time-intervals will be treble; and that in four, quadruple that of the first time-interval. To put the matter more clearly, if a body were to continue its motion with the same speed which it had acquired during the first time-interval and were to retain this same uniform speed, then its motion would be twice as slow as that which it would have if its velocity had been acquired during *two* time-intervals.

And thus, it seems, we shall not be far wrong if we put the increment of speed as proportional to the increment of time; hence

hence the definition of motion which we are about to discuss may be stated as follows: A motion is said to be uniformly accelerated, when starting from rest, it acquires, during equal time-intervals, equal increments of speed.

SAGR. Although I can offer no rational objection to this or indeed to any other definition, devised by any author whomsoever, since all definitions are arbitrary, I may nevertheless without offense be allowed to doubt whether such a definition as the above, established in an abstract manner, corresponds to and describes that kind of accelerated motion which we meet in nature in the case of freely falling bodies. And since the Author apparently maintains that the motion described in his definition is that of freely falling bodies, I would like to clear my mind of certain difficulties in order that I may later apply myself more earnestly to the propositions and their demonstrations.

SALV. It is well that you and Simplicio raise these difficulties. They are, I imagine, the same which occurred to me when I first saw this treatise, and which were removed either by discussion with the Author himself, or by turning the matter over in my own mind.

SAGR. When I think of a heavy body falling from rest, that is, starting with zero speed and gaining speed in proportion to the

[199]

time from the beginning of the motion; such a motion as would, for instance, in eight beats of the pulse acquire eight degrees of speed; having at the end of the fourth beat acquired four degrees; at the end of the second, two; at the end of the first, one: and since time is divisible without limit, it follows from all these considerations that if the earlier speed of a body is less than its present speed in a constant ratio, then there is no degree of speed however small (or, one may say, no degree of slowness however great) with which we may not find this body travelling after starting from infinite slowness, i. e., from rest. So that if that speed which it had at the end of the fourth beat was such that, if kept uniform, the body would traverse two miles in an hour, and if keeping the speed which it had at the end of the second

second beat, it would traverse one mile an hour, we must infer that, as the instant of starting is more and more nearly approached, the body moves so slowly that, if it kept on moving at this rate, it would not traverse a mile in an hour, or in a day, or in a year or in a thousand years; indeed, it would not traverse a span in an even greater time; a phenomenon which baffles the imagination, while our senses show us that a heavy falling body suddenly acquires great speed.

SALV. This is one of the difficulties which I also at the beginning, experienced, but which I shortly afterwards removed; and the removal was effected by the very experiment which creates the difficulty for you. You say the experiment appears to show that immediately after a heavy body starts from rest it acquires a very considerable speed: and I say that the same experiment makes clear the fact that the initial motions of a falling body, no matter how heavy, are very slow and gentle. Place a heavy body upon a yielding material, and leave it there without any pressure except that owing to its own weight; it is clear that if one lifts this body a cubit or two and allows it to fall upon the same material, it will, with this impulse, exert a new and greater pressure than that caused by its mere weight; and this effect is brought about by the [weight of the] falling body together with the velocity acquired during the fall, an effect which will be greater and greater according to the height of the fall, that is according as the velocity of the falling body becomes greater. From the quality and intensity of the blow we are thus enabled to accurately estimate the speed of a falling body. But tell me, gentlemen, is it not true that if a block be allowed to fall upon a stake from a height of four cubits and drives it into the earth,

[200]

say, four finger-breadths, that coming from a height of two cubits it will drive the stake a much less distance, and from the height of one cubit a still less distance; and finally if the block be lifted only one finger-breadth how much more will it accomplish than if merely laid on top of the stake without percussion? Certainly very little. If it be lifted only the thickness of a leaf, the effect will be altogether imperceptible. And since the
effect

222

effect of the blow depends upon the velocity of this striking body, can any one doubt the motion is very slow and the speed more than small whenever the effect [of the blow] is imperceptible? See now the power of truth; the same experiment which at first glance seemed to show one thing, when more carefully examined, assures us of the contrary.

But without depending upon the above experiment, which is doubtless very conclusive, it seems to me that it ought not to be difficult to establish such a fact by reasoning alone. Imagine a heavy stone held in the air at rest; the support is removed and the stone set free; then since it is heavier than the air it begins to fall, and not with uniform motion but slowly at the beginning and with a continuously accelerated motion. Now since velocity can be increased and diminished without limit, what reason is there to believe that such a moving body starting with infinite slowness, that is, from rest, immediately acquires a speed of ten degrees rather than one of four, or of two, or of one, or of a half, or of a hundredth; or, indeed, of any of the infinite number of small values [of speed]? Pray listen. I hardly think you will refuse to grant that the gain of speed of the stone falling from rest follows the same sequence as the diminution and loss of this same speed when, by some impelling force, the stone is thrown to its former elevation: but even if you do not grant this, I do not see how you can doubt that the ascending stone, diminishing in speed, must before coming to rest pass through every possible degree of slowness.

SIMP. But if the number of degrees of greater and greater slowness is limitless, they will never be all exhausted, therefore such an ascending heavy body will never reach rest, but will continue to move without limit always at a slower rate; but this is not the observed fact.

SALV. This would happen, Simplicio, if the moving body were to maintain its speed for any length of time at each degree of velocity; but it merely passes each point without delaying more than an instant: and since each time-interval however

[201]

small may be divided into an infinite number of instants, these will

223

will always be sufficient [in number] to correspond to the infinite degrees of diminished velocity.

That such a heavy rising body does not remain for any length of time at any given degree of velocity is evident from the following: because if, some time-interval having been assigned, the body moves with the same speed in the last as in the first instant of that time-interval, it could from this second degree of elevation be in like manner raised through an equal height, just as it was transferred from the first elevation to the second, and by the same reasoning would pass from the second to the third and would finally continue in uniform motion forever.

SAGR. From these considerations it appears to me that we may obtain a proper solution of the problem discussed by philosophers, namely, what causes the acceleration in the natural motion of heavy bodies? Since, as it seems to me, the force [*virtù*] impressed by the agent projecting the body upwards diminishes continuously, this force, so long as it was greater than the contrary force of gravitation, impelled the body upwards; when the two are in equilibrium the body ceases to rise and passes through the state of rest in which the impressed impetus [*impeto*] is not destroyed, but only its excess over the weight of the body has been consumed—the excess which caused the body to rise. Then as the diminution of the outside impetus [*impeto*] continues, and gravitation gains the upper hand, the fall begins, but slowly at first on account of the opposing impetus [*virtù impressa*], a large portion of which still remains in the body; but as this continues to diminish it also continues to be more and more overcome by gravity, hence the continuous acceleration of motion.

SIMP. The idea is clever, yet more subtle than sound; for even if the argument were conclusive, it would explain only the case in which a natural motion is preceded by a violent motion, in which there still remains active a portion of the external force [*virtù esterna*]; but where there is no such remaining portion and the body starts from an antecedent state of rest, the cogency of the whole argument fails.

SAGR. I believe that you are mistaken and that this distinction

tion between cases which you make is superfluous or rather non-existent. But, tell me, cannot a projectile receive from the projector either a large or a small force [*virtù*] such as will throw it to a height of a hundred cubits, and even twenty or four or one?

[202]

SIMP. Undoubtedly, yes.

SAGR. So therefore this impressed force [*virtù impressa*] may exceed the resistance of gravity so slightly as to raise it only a finger-breadth; and finally the force [*virtù*] of the projector may be just large enough to exactly balance the resistance of gravity so that the body is not lifted at all but merely sustained. When one holds a stone in his hand does he do anything but give it a force impelling [*virtù impellente*] it upwards equal to the power [*facoltà*] of gravity drawing it downwards? And do you not continuously impress this force [*virtù*] upon the stone as long as you hold it in the hand? Does it perhaps diminish with the time during which one holds the stone?

And what does it matter whether this support which prevents the stone from falling is furnished by one's hand or by a table or by a rope from which it hangs? Certainly nothing at all. You must conclude, therefore, Simplicio, that it makes no difference whatever whether the fall of the stone is preceded by a period of rest which is long, short, or instantaneous provided only the fall does not take place so long as the stone is acted upon by a force [*virtù*] opposed to its weight and sufficient to hold it at rest.

SALV. The present does not seem to be the proper time to investigate the cause of the acceleration of natural motion concerning which various opinions have been expressed by various philosophers, some explaining it by attraction to the center, others to repulsion between the very small parts of the body, while still others attribute it to a certain stress in the surrounding medium which closes in behind the falling body and drives it from one of its positions to another. Now, all these fantasies, and others too, ought to be examined; but it is not really worth while. At present it is the purpose of our Author merely to investigate

investigate and to demonstrate some of the properties of accelerated motion (whatever the cause of this acceleration may be)—meaning thereby a motion, such that the momentum of its velocity [*i momenti della sua velocità*] goes on increasing after departure from rest, in simple proportionality to the time, which is the same as saying that in equal time-intervals the body receives equal increments of velocity; and if we find the properties [of accelerated motion] which will be demonstrated later are realized in freely falling and accelerated bodies, we may conclude that the assumed definition includes such a motion of falling bodies and that their speed [*accelerazione*] goes on increasing as the time and the duration of the motion.

[203]

SAGR. So far as I see at present, the definition might have been put a little more clearly perhaps without changing the fundamental idea, namely, uniformly accelerated motion is such that its speed increases in proportion to the space traversed; so that, for example, the speed acquired by a body in falling four cubits would be double that acquired in falling two cubits and this latter speed would be double that acquired in the first cubit. Because there is no doubt but that a heavy body falling from the height of six cubits has, and strikes with, a momentum [*impeto*] double that it had at the end of three cubits, triple that which it had at the end of one.

SALV. It is very comforting to me to have had such a companion in error; and moreover let me tell you that your proposition seems so highly probable that our Author himself admitted, when I advanced this opinion to him, that he had for some time shared the same fallacy. But what most surprised me was to see two propositions so inherently probable that they commanded the assent of everyone to whom they were presented, proven in a few simple words to be not only false, but impossible.

SIMP. I am one of those who accept the proposition, and believe that a falling body acquires force [*vires*] in its descent, its velocity increasing in proportion to the space, and that the momentum [*momento*] of the falling body is doubled when it falls
from

from a doubled height; these propositions, it appears to me, ought to be conceded without hesitation or controversy.

SALV. And yet they are as false and impossible as that motion should be completed instantaneously; and here is a very clear demonstration of it. If the velocities are in proportion to the spaces traversed, or to be traversed, then these spaces are traversed in equal intervals of time; if, therefore, the velocity with which the falling body traverses a space of eight feet were double that with which it covered the first four feet (just as the one distance is double the other) then the time-intervals required for these passages would be equal. But for one and the same body to fall eight feet and four feet in the same time is possible only in the case of instantaneous [discontinuous] motion;

[204]

but observation shows us that the motion of a falling body occupies time, and less of it in covering a distance of four feet than of eight feet; therefore it is not true that its velocity increases in proportion to the space.

The falsity of the other proposition may be shown with equal clearness. For if we consider a single striking body the difference of momentum in its blows can depend only upon difference of velocity; for if the striking body falling from a double height were to deliver a blow of double momentum, it would be necessary for this body to strike with a doubled velocity; but with this doubled speed it would traverse a doubled space in the same time-interval; observation however shows that the time required for fall from the greater height is longer.

SAGR. You present these recondite matters with too much evidence and ease; this great facility makes them less appreciated than they would be had they been presented in a more abstruse manner. For, in my opinion, people esteem more lightly that knowledge which they acquire with so little labor than that acquired through long and obscure discussion.

SALV. If those who demonstrate with brevity and clearness the fallacy of many popular beliefs were treated with contempt instead of gratitude the injury would be quite bearable; but on the other hand it is very unpleasant and annoying to see men, who

227

who claim to be peers of anyone in a certain field of study, take for granted certain conclusions which later are quickly and easily shown by another to be false. I do not describe such a feeling as one of envy, which usually degenerates into hatred and anger against those who discover such fallacies; I would call it a strong desire to maintain old errors, rather than accept newly discovered truths. This desire at times induces them to unite against these truths, although at heart believing in them, merely for the purpose of lowering the esteem in which certain others are held by the unthinking crowd. Indeed, I have heard from our Academician many such fallacies held as true but easily refutable; some of these I have in mind.

SAGR. You must not withhold them from us, but, at the proper time, tell us about them even though an extra session be necessary. But now, continuing the thread of our talk, it would

[205]

seem that up to the present we have established the definition of uniformly accelerated motion which is expressed as follows:

A motion is said to be equally or uniformly accelerated when, starting from rest, its momentum (*celeritatis momenta*) receives equal increments in equal times.

SALV. This definition established, the Author makes a single assumption, namely,

The speeds acquired by one and the same body moving down planes of different inclinations are equal when the heights of these planes are equal.

By the height of an inclined plane we mean the perpendicular let fall from the upper end of the plane upon the horizontal line drawn through the lower end of the same plane. Thus, to illustrate, let the line AB be horizontal, and let the planes CA and CD be inclined to it; then the Author calls the perpendicular CB the "height" of the planes CA and CD; he supposes that the speeds acquired by one and the same body, descending along the planes CA and CD to the terminal points A and D are equal since the heights of these planes are the same, CB; and also it must be understood that this speed is that which would be acquired by the same body falling from C to B.

Sagr.

SAGR. Your assumption appears to me so reasonable that it ought to be conceded without question, provided of course there are no chance or outside resistances, and that the planes are hard and smooth, and that the figure of the moving body is perfectly round, so that neither plane nor moving body is rough. All resistance and opposition having been removed, my reason tells me at once that a heavy and perfectly round ball descending along the lines CA, CD, CB would reach the terminal points A, D, B, with equal momenta [*impeti eguali*].

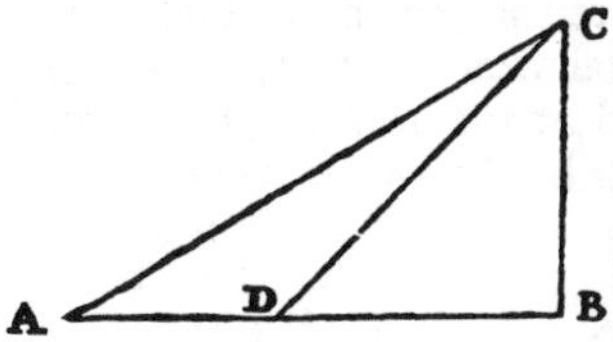

Fig. 45

SALV. Your words are very plausible; but I hope by experiment to increase the probability to an extent which shall be little short of a rigid demonstration.

[206]

Imagine this page to represent a vertical wall, with a nail driven into it; and from the nail let there be suspended a lead bullet of one or two ounces by means of a fine vertical thread, AB, say from four to six feet long, on this wall draw a horizontal line DC, at right angles to the vertical thread AB, which hangs about two finger-breadths in front of the wall. Now bring the thread AB with the attached ball into the position AC and set it free; first it will be observed to descend along the arc CBD, to pass the point B, and to travel along the arc BD, till it almost reaches the horizontal CD, a slight shortage being caused by the resistance of the air and the string; from this we may rightly infer that the ball in its descent through the arc CB acquired a momentum [*impeto*] on reaching B, which was just sufficient to carry it through a similar arc BD to the same height. Having repeated this experiment many times, let us now drive a nail into the wall close to the perpendicular AB, say at E or F, so that it projects out some five or six finger-breadths in order that the thread, again carrying the bullet through the arc CB, may strike upon the nail E when the bullet reaches B, and thus compel it to traverse the arc BG, described about E as center. From this we

229

we can see what can be done by the same momentum [*impeto*] which previously starting at the same point B carried the same body through the arc BD to the horizontal CD. Now, gentlemen, you will observe with pleasure that the ball swings to the point G in the horizontal, and you would see the same thing happen if the obstacle were placed at some lower point, say at F, about which the ball would describe the arc BI, the rise of the

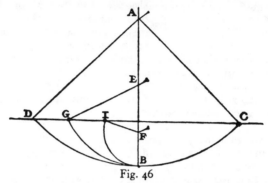

Fig. 46

ball always terminating exactly on the line CD. But when the nail is placed so low that the remainder of the thread below it will not reach to the height CD (which would happen if the nail were placed nearer B than to the intersection of AB with the

[207]

horizontal CD) then the thread leaps over the nail and twists itself about it.

This experiment leaves no room for doubt as to the truth of our supposition; for since the two arcs CB and DB are equal and similarly placed, the momentum [*momento*] acquired by the fall through the arc CB is the same as that gained by fall through the arc DB; but the momentum [*momento*] acquired at B, owing to fall through CB, is able to lift the same body [*mobile*] through the arc BD; therefore, the momentum acquired in the fall BD is equal to that which lifts the same body through the same arc from B to D; so, in general, every momentum acquired by fall through

through an arc is equal to that which can lift the same body through the same arc. But all these momenta [*momenti*] which cause a rise through the arcs BD, BG, and BI are equal, since they are produced by the same momentum, gained by fall through CB, as experiment shows. Therefore all the momenta gained by fall through the arcs DB, GB, IB are equal.

SAGR. The argument seems to me so conclusive and the experiment so well adapted to establish the hypothesis that we may, indeed, consider it as demonstrated.

SALV. I do not wish, Sagredo, that we trouble ourselves too much about this matter, since we are going to apply this principle mainly in motions which occur on plane surfaces, and not upon curved, along which acceleration varies in a manner greatly different from that which we have assumed for planes.

So that, although the above experiment shows us that the descent of the moving body through the arc CB confers upon it momentum [*momento*] just sufficient to carry it to the same height through any of the arcs BD, BG, BI, we are not able, by similar means, to show that the event would be identical in the case of a perfectly round ball descending along planes whose inclinations are respectively the same as the chords of these arcs. It seems likely, on the other hand, that, since these planes form angles at the point B, they will present an obstacle to the ball which has descended along the chord CB, and starts to rise along the chord BD, BG, BI.

In striking these planes some of its momentum [*impeto*] will be lost and it will not be able to rise to the height of the line CD; but this obstacle, which interferes with the experiment, once removed, it is clear that the momentum [*impeto*] (which gains

[208]

in strength with descent) will be able to carry the body to the same height. Let us then, for the present, take this as a postulate, the absolute truth of which will be established when we find that the inferences from it correspond to and agree perfectly with experiment. The author having assumed this single principle passes next to the propositions which he clearly demonstrates; the first of these is as follows:

THIRD DAY

The time in which any space is traversed by a body starting from rest and uniformly accelerated is equal to the time in which that same space would be traversed by the same body moving at a uniform speed whose value is the mean of the highest speed and the speed just before acceleration began.

Let us represent by the line AB the time in which the space CD is traversed by a body which starts from rest at C and is uniformly accelerated; let the final and highest value of the speed gained during the interval AB be represented by the line EB drawn at right angles to AB; draw the line AE, then all lines drawn from equidistant points on AB and parallel to BE will represent the increasing values of the speed, beginning with the instant A. Let the point F bisect the line EB; draw FG parallel to BA, and GA parallel to FB, thus forming a parallelogram AGFB which will be equal in area to the triangle AEB, since the side GF bisects the side AE at the point I; for if the parallel lines in the triangle AEB are extended to GI, then the sum of all the parallels contained in the quadrilateral is equal to the sum of those contained in the triangle AEB; for those in the triangle IEF are equal to those contained in the triangle GIA, while those included in the trapezium AIFB are common. Since each and every instant of time in the time-interval AB has its corresponding point on the line AB, from which points parallels drawn in and limited by the triangle AEB represent the increasing values of the growing velocity, and since parallels contained within the rectangle represent the values of a speed which is not increasing, but constant, it appears, in like manner, that the momenta [*momenta*] assumed by the moving body may also be represented, in the case of the accelerated motion, by the increasing parallels of the triangle AEB,

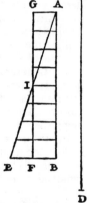

Fig. 47

AEB, and, in the case of the uniform motion, by the parallels of the rectangle GB. For, what the momenta may lack in the first part of the accelerated motion (the deficiency of the momenta being represented by the parallels of the triangle AGI) is made up by the momenta represented by the parallels of the triangle IEF.

Hence it is clear that equal spaces will be traversed in equal times by two bodies, one of which, starting from rest, moves with a uniform acceleration, while the momentum of the other, moving with uniform speed, is one-half its maximum momentum under accelerated motion.   Q. E. D.

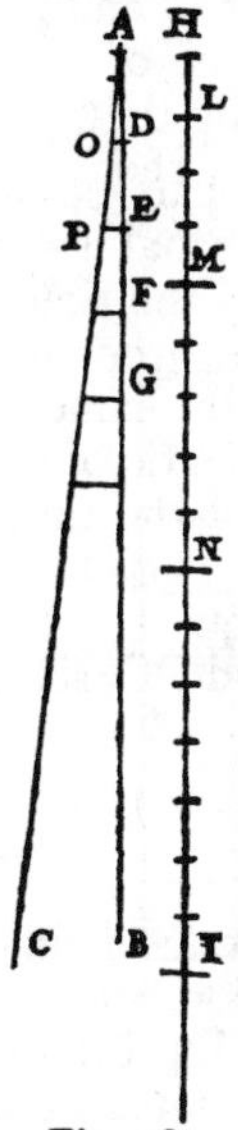

THEOREM II, PROPOSITION II

The spaces described by a body falling from rest with a uniformly accelerated motion are to each other as the squares of the time-intervals employed in traversing these distances.

Let the time beginning with any instant A be represented by the straight line AB in which are taken any two time-intervals AD and AE. Let HI represent the distance through which the body, starting from rest at H, falls with uniform acceleration. If HL represents the space traversed during the time-interval AD, and HM that covered during the interval AE, then the space MH stands to the space LH in a ratio which is the square of the ratio of the time AE to the time AD; or we may say simply that the distances HM and HL are related as the squares of AE and AD.

Fig. 48

Draw the line AC making any angle whatever with the line AB; and from the points D and E, draw the parallel lines DO and EP; of these two lines, DO represents the greatest velocity attained during the interval AD, while EP represents the maximum velocity acquired during the interval AE. But it has just been proved that so far as distances traversed are concerned

cerned it is precisely the same whether a body falls from rest with a uniform acceleration or whether it falls during an equal time-interval with a constant speed which is one-half the maximum speed attained during the accelerated motion. It follows therefore that the distances HM and HL are the same as would be traversed, during the time-intervals AE and AD, by uniform velocities equal to one-half those represented by DO and EP respectively. If, therefore, one can show that the distances HM and HL are in the same ratio as the squares of the time-intervals AE and AD, our proposition will be proven.

[210]

But in the fourth proposition of the first book [p. 157 above] it has been shown that the spaces traversed by two particles in uniform motion bear to one another a ratio which is equal to the product of the ratio of the velocities by the ratio of the times. But in this case the ratio of the velocities is the same as the ratio of the time-intervals (for the ratio of AE to AD is the same as that of ½ EP to ½ DO or of EP to DO). Hence the ratio of the spaces traversed is the same as the squared ratio of the time-intervals.   Q. E. D.

Evidently then the ratio of the distances is the square of the ratio of the final velocities, that is, of the lines EP and DO, since these are to each other as AE to AD.

COROLLARY I

Hence it is clear that if we take any equal intervals of time whatever, counting from the beginning of the motion, such as AD, DE, EF, FG, in which the spaces HL, LM, MN, NI are traversed, these spaces will bear to one another the same ratio as the series of odd numbers, 1, 3, 5, 7; for this is the ratio of the differences of the squares of the lines [which represent time], differences which exceed one another by equal amounts, this excess being equal to the smallest line [viz. the one representing a single time-interval]: or we may say [that this is the ratio] of the differences of the squares of the natural numbers beginning with unity.

While,

While, therefore, during equal intervals of time the velocities increase as the natural numbers, the increments in the distances traversed during these equal time-intervals are to one another as the odd numbers beginning with unity.

SAGR. Please suspend the discussion for a moment since there just occurs to me an idea which I want to illustrate by means of a diagram in order that it may be clearer both to you and to me.

Let the line AI represent the lapse of time measured from the initial instant A; through A draw the straight line AF making any angle whatever; join the terminal points I and F; divide the time AI in half at C; draw CB parallel to IF. Let us consider CB as the maximum value of the velocity which increases from zero at the beginning, in simple proportionality to the intercepts on the triangle ABC of lines drawn parallel to BC; or what is the same thing, let us suppose the velocity to increase in proportion to the time; then I admit without question, in view of the preceding argument, that the space described by a body falling in the aforesaid manner will be equal to the space traversed by the same body during the same length of time travelling with a uniform speed equal to EC, the half of BC. Further let us imagine that the body has fallen with accelerated motion so that, at the instant C, it has the velocity BC. It is clear that if the body continued to descend with the same speed BC, without acceleration, it would in the next time-interval CI traverse double the distance covered during the interval AC, with the uniform speed EC which is half of BC; but since the falling body acquires equal increments of speed during equal increments of time, it follows that the velocity BC, during the next time-interval

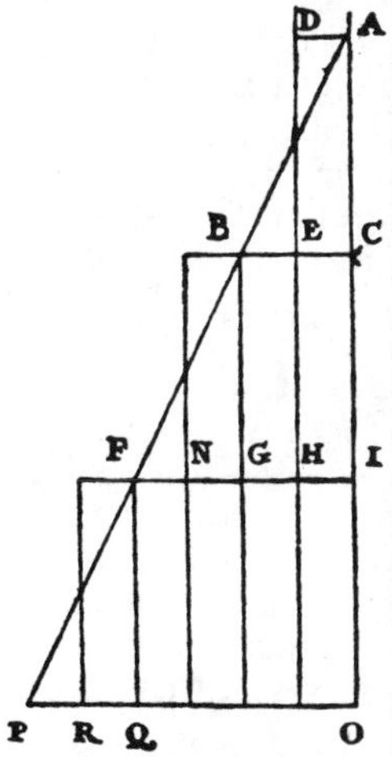

Fig. 49   [211]

interval

235

interval CI will be increased by an amount represented by the parallels of the triangle BFG which is equal to the triangle ABC. If, then, one adds to the velocity GI half of the velocity FG, the highest speed acquired by the accelerated motion and determined by the parallels of the triangle BFG, he will have the uniform velocity with which the same space would have been described in the time CI; and since this speed IN is three times as great as EC it follows that the space described during the interval CI is three times as great as that described during the interval AC. Let us imagine the motion extended over another equal time-interval IO, and the triangle extended to APO; it is then evident that if the motion continues during the interval IO, at the constant rate IF acquired by acceleration during the time AI, the space traversed during the interval IO will be four times that traversed during the first interval AC, because the speed IF is four times the speed EC. But if we enlarge our triangle so as to include FPQ which is equal to ABC, still assuming the acceleration to be constant, we shall add to the uniform speed an increment RQ, equal to EC; then the value of the equivalent uniform speed during the time-interval IO will be five times that during the first time-interval AC; therefore the space traversed will be quintuple that during the first interval AC. It is thus evident by simple computation that a moving body starting from rest and acquiring velocity at a rate proportional to the time, will, during equal intervals of time, traverse distances which are related to each other as the odd numbers beginning with unity, 1, 3, 5; [*] or considering the total space traversed, that covered

[212[

in double time will be quadruple that covered during unit time; in triple time, the space is nine times as great as in unit time.

---

[*] As illustrating the greater elegance and brevity of modern analytical methods, one may obtain the result of Prop. II directly from the fundamental equation

$$s = \tfrac{1}{2} g \, (t^2_2 - t^2_1) = g/2 \, (t_2 + t_1) \, (t_2 - t_1)$$

where g is the acceleration of gravity and s, the space traversed between the instants t_1 and t_2. If now $t_2 - t_1 = 1$, say one second, then $s = g/2 \, (t_2 + t_1)$ where $t_2 + t_1$, must always be an odd number, seeing that it is the sum of two consecutive terms in the series of natural numbers. [*Trans.*]

And in general the spaces traversed are in the duplicate ratio of the times, i. e., in the ratio of the squares of the times.

SIMP. In truth, I find more pleasure in this simple and clear argument of Sagredo than in the Author's demonstration which to me appears rather obscure; so that I am convinced that matters are as described, once having accepted the definition of uniformly accelerated motion. But as to whether this acceleration is that which one meets in nature in the case of falling bodies, I am still doubtful; and it seems to me, not only for my own sake but also for all those who think as I do, that this would be the proper moment to introduce one of those experiments—and there are many of them, I understand—which illustrate in several ways the conclusions reached.

SALV. The request which you, as a man of science, make, is a very reasonable one; for this is the custom—and properly so—in those sciences where mathematical demonstrations are applied to natural phenomena, as is seen in the case of perspective, astronomy, mechanics, music, and others where the principles, once established by well-chosen experiments, become the foundations of the entire superstructure. I hope therefore it will not appear to be a waste of time if we discuss at considerable length this first and most fundamental question upon which hinge numerous consequences of which we have in this book only a small number, placed there by the Author, who has done so much to open a pathway hitherto closed to minds of speculative turn. So far as experiments go they have not been neglected by the Author; and often, in his company, I have attempted in the following manner to assure myself that the acceleration actually experienced by falling bodies is that above described.

A piece of wooden moulding or scantling, about 12 cubits long, half a cubit wide, and three finger-breadths thick, was taken; on its edge was cut a channel a little more than one finger in breadth; having made this groove very straight, smooth, and polished, and having lined it with parchment, also as smooth and polished as possible, we rolled along it a hard, smooth, and very round bronze ball. Having placed this

board

board in a sloping position, by lifting one end some one or two cubits above the other, we rolled the ball, as I was just saying, along the channel, noting, in a manner presently to be described, the time required to make the descent. We repeated this experiment more than once in order to measure the time with an accuracy such that the deviation between two observations never exceeded one-tenth of a pulse-beat. Having performed this operation and having assured ourselves of its reliability, we now rolled the ball only one-quarter the length of the channel; and having measured the time of its descent, we found it precisely one-half of the former. Next we tried other distances, comparing the time for the whole length with that for the half, or with that for two-thirds, or three-fourths, or indeed for any fraction; in such experiments, repeated a full hundred times, we always found that the spaces traversed were to each other as the squares of the times, and this was true for all inclinations of the plane, i. e., of the channel, along which we rolled the ball. We also observed that the times of descent, for various inclinations of the plane, bore to one another precisely that ratio which, as we shall see later, the Author had predicted and demonstrated for them.

For the measurement of time, we employed a large vessel of water placed in an elevated position; to the bottom of this vessel was soldered a pipe of small diameter giving a thin jet of water, which we collected in a small glass during the time of each descent, whether for the whole length of the channel or for a part of its length; the water thus collected was weighed, after each descent, on a very accurate balance; the differences and ratios of these weights gave us the differences and ratios of the times, and this with such accuracy that although the operation was repeated many, many times, there was no appreciable discrepancy in the results.

SIMP. I would like to have been present at these experiments; but feeling confidence in the care with which you performed them, and in the fidelity with which you relate them, I am satisfied and accept them as true and valid

SALV. Then we can proceed without discussion.

In some respects Sir Isaac Newton (1642-1727) had the most original mind of modern times. The mathematician and natural philosopher achieved a synthesis of ideas which became the cornerstone of the scientific revolution. Newton was the first to bring the full power of mathematical analysis to bear on the problems of a science of motion.

PRINCIPIA MATHEMATICAS

PREFACE TO THE FIRST EDITION

SINCE the ancients (as we are told by Pappus) esteemed the science of mechanics of greatest importance in the investigation of natural things, and the moderns, rejecting substantial forms and occult qualities, have endeavored to subject the phenomena of nature to the laws of mathematics, I have in this treatise cultivated mathematics as far as it relates to philosophy. The ancients considered mechanics in a twofold respect; as rational, which proceeds accurately by demonstration, and practical. To practical mechanics all the manual arts belong, from which mechanics took its name. But as artificers do not work with perfect accuracy, it comes to pass that mechanics is so distinguished from geometry that what is perfectly accurate is called geometrical; what is less so, is called mechanical. However, the errors are not in the art, but in the artificers. He that works with less accuracy is an imperfect mechanic; and if any could work with perfect accuracy, he would be the most perfect mechanic of all, for the description of right lines and circles, upon which geometry is founded, belongs to mechanics. Geometry does not teach us to draw these lines, but requires them to be drawn, for it requires that the learner should first be taught to describe these accurately before he enters upon geometry, then it shows how by these operations problems may be solved. To describe right lines and circles are problems, but not geometrical problems. The solution of these problems is required from mechanics, and by geometry the use of them, when so solved, is shown; and it is the glory of geometry that from those few principles, brought from without, it is able to produce so many things. Therefore geometry is founded in mechanical practice, and is nothing but that part of universal mechanics which accurately proposes and demonstrates the art of measuring. But since the manual arts are chiefly employed in the moving of bodies, it happens that geometry is commonly referred to their magnitude, and mechanics to their motion. In this sense rational mechanics will be the science of motions resulting from any forces whatsoever, and of the forces required to produce any motions, accurately proposed and demonstrated. This part of mechanics, as far as it extended to the five powers which relate to manual arts, was cultivated by the ancients, who considered gravity (it not being a manual power) no otherwise than in moving weights by those powers. But I consider philosophy rather than arts and write not concerning manual but natural powers, and consider chiefly those things which relate to gravity, levity, elastic force, the resistance of fluids, and the like forces, whether attractive or impulsive; and therefore I offer this work as the mathematical principles of philosophy, for the whole burden of philosophy seems to consist in this—from the phenomena of motions to investigate the forces of nature, and then from these forces to demonstrate the other phenomena; and to this end the general propositions in the first and second books are directed. In the third book I give an example of this in the explication of the System of the World; for by the propositions mathematically demonstrated in the former books in the third I derive from the celestial phenomena the forces of gravity with which bodies tend to the sun and the several planets. Then from these forces, by other propositions which are also mathematical, I deduce the motions of the planets, the comets, the moon, and the sea. I wish we could derive the rest of the phenomena of Nature by the same kind of reasoning from mechanical principles, for I am induced by many reasons to suspect that they may all depend upon certain forces by which the particles of bodies, by some causes hitherto unknown, are either mutually impelled towards one another, and cohere in regular figures, or are repelled and recede from one another. These forces being unknown, philosophers have hitherto attempted the search of Nature in vain; but I hope the principles here laid down will afford some light either to this or some truer method of philosophy.

In the publication of this work the most acute and universally learned Mr. Edmund Halley not only assisted me in correcting the errors of the press and preparing the geometrical figures, but it was through his solicitations that it came to be published; for when he had obtained of me my demonstrations of the figure of the celestial orbits, he continually pressed me to communicate the same to the Royal Society, who afterwards, by their kind encouragement and entreaties, engaged me to think of publishing them. But after I had begun to consider the inequalities of the lunar motions, and had entered upon some other things relating to the laws and measures of gravity and other forces; and the figures that would be described by bodies attracted according to given laws; and the motion of several bodies moving among themselves; the motion of bodies in resisting mediums; the forces, densities, and motions, of mediums; the orbits of the comets, and such like, I deferred that publication till I had made a search into those matters, and could put forth the whole together. What relates to the lunar motions (being imperfect), I have put all together in the corollaries of Prop. 66, to avoid being obliged to propose and distinctly demonstrate the several things there contained in a method more prolix than the subject deserved and interrupt the series of the other propositions. Some things, found out after the rest, I chose to insert in places less suitable, rather than change the number of the propositions and the citations. I heartily beg that what I have here done may be read with forbearance; and that my labors in a subject so difficult may be examined, not so much with the view to censure, as to remedy their defects.

Is. NEWTON

Cambridge, Trinity College, *May* 8, 1686

Hitherto I have laid down the definitions of such words as are less known, and explained the sense in which I would have them to be understood in the following discourse. I do not define time, space, place, and motion, as being well known to all. Only I must observe, that the common people conceive those quantities under no other notions but from the relation they bear to sensible objects. And thence arise certain prejudices, for the removing of which it will be convenient to distinguish them into absolute and relative, true and apparent, mathematical and common.

I. Absolute, true, and mathematical time, of itself, and from its own nature, flows equably without relation to anything external, and by another name is called duration: relative, apparent, and common time, is some sensible and external (whether accurate or unequable) measure of duration by the means of motion, which is commonly used instead of true time; such as an hour, a day, a month, a year.

II. Absolute space, in its own nature, without relation to anything external, remains always similar and immovable. Relative space is some movable dimension or measure of the absolute spaces; which our senses determine by its position to bodies; and which is commonly taken for immovable space; such is the dimension of a subterraneous, an aerial, or celestial space, determined by its position in respect of the earth. Absolute and relative space are the same in figure and magnitude; but they do not remain always numerically the same.

For if the earth, for instance, moves, a space of our air, which relatively and in respect of the earth remains always the same, will at one time be one part of the absolute space into which the air passes; at another time it will be another part of the same, and so, absolutely understood, it will be continually changed.

III. Place is a part of space which a body takes up, and is according to the space, either absolute or relative. I say, a part of space; not the situation, nor the external surface of the body. For the places of equal solids are always equal; but their surfaces, by reason of their dissimilar figures, are often unequal. Positions properly have no quantity, nor are they so much the places themselves, as the properties of places. The motion of the whole is the same with the sum of the motions of the parts; that is, the translation of the whole, out of its place, is the same thing with the sum of the translations of the parts out of their places; and therefore the place of the whole is the same as the sum of the places of the parts, and for that reason, it is internal, and in the whole body.

IV. Absolute motion is the translation of a body from one absolute place into another; and relative motion, the translation from one relative place into another. Thus in a ship under sail, the relative place of a body is that part of the ship which the body possesses; or that part of the cavity which the body fills, and which therefore moves together with the ship: and relative rest is the continuance of the body in the same part of the ship, or of its cavity. But real, absolute rest, is the continuance of the body in the same part of that immovable space, in which the ship itself, its cavity, and all that it contains, is moved. Wherefore, if the earth is really at rest, the body, which relatively rests in the ship, will really and absolutely move with the same velocity which the ship has on the earth. But if the earth also moves, the true and absolute motion of the body will arise, partly from the true motion of the earth, in immovable space, partly from the relative motion of the ship on the earth; and if the body moves also relatively in the ship, its true motion will arise, partly from the true motion of the earth, in immovable space, and partly from the relative motions as well of the ship on the earth, as of the body in the ship; and from these relative motions will arise the relative motion of the body on the earth. As if that part of the earth, where the ship is, was truly moved towards the east, with a velocity of 10,010 parts; while the ship itself, with a fresh gale, and full sails, is carried towards the west, with a velocity expressed by 10 of those parts; but a sailor walks in the ship towards the east, with 1 part of the said velocity; then the sailor will be moved truly in immovable space towards the east, with a velocity of 10,001 parts, and relatively on the earth towards the west, with a velocity of 9 of those parts.

Absolute time, in astronomy, is distinguished from relative, by the equation or correction of the apparent time. For the natural days are truly unequal, though they are commonly considered as equal, and used for a measure of time; astronomers correct this inequality that they may measure the celestial motions by a more accurate time. It may be, that there is no such thing as an equable motion, whereby time may be accurately measured. All motions may be accelerated and retarded, but the flowing of absolute time is not liable to any change. The duration of perseverance of the existence of things remains the same, whether the motions are swift or slow, or none at all: and therefore this duration ought to be distinguished from what are only sensible measures thereof; and from which we deduce it, by means of the astronomical equation. The

243

necessity of this equation, for determining the times of a phenomenon, is evinced as well from the experiments of the pendulum clock, as by eclipses of the satellites of Jupiter.

As the order of the parts of time is immutable, so also is the order of the parts of space. Suppose those parts to be moved out of their places, and they will be moved (if the expression may be allowed) out of themselves. For times and spaces are, as it were, the places as well of themselves as of all other things. All things are placed in time as to order of succession; and in space as to order of situation. It is from their essence or nature that they are places; and that the primary places of things should be movable, is absurd. These are therefore the absolute places; and translations out of those places, are the only absolute motions.

But because the parts of space cannot be seen, or distinguished from one another by our senses, therefore in their stead we use sensible measures of them. For from the positions and distances of things from any body considered as immovable, we define all places; and then with respect to such places, we estimate all motions, considering bodies as transferred from some of those places into others. And so, instead of absolute places and motions, we use relative ones; and that without any inconvenience in common affairs; but in philosophical disquisitions, we ought to abstract from our senses, and consider things themselves, distinct from what are only sensible measures of them. For it may be that there is no body really at rest, to which the places and motions of others may be referred.

But we may distinguish rest and motion, absolute and relative, one from the other by their properties, causes, and effects. It is a property of rest, that bodies really at rest do rest in respect to one another. And therefore as it is possible, that in the remote regions of the fixed stars, or perhaps far beyond them, there may be some body absolutely at rest; but impossible to know, from the position of bodies to one another in our regions, whether any of these do keep the same position to that remote body, it follows that absolute rest cannot be determined from the position of bodies in our regions.

It is a property of motion, that the parts, which retain given positions to their wholes, do partake of the motions of those wholes. For all the parts of revolving bodies endeavor to recede from the axis of motion; and the impetus of bodies moving forwards arises from the joint impetus of all the parts. Therefore, if surrounding bodies are moved, those that are relatively at rest within them will partake of their motion. Upon which account, the true and absolute motion of a body cannot be determined by the translation of it from those which only seem to rest; for the external bodies ought not only to appear at rest, but to be really at rest. For otherwise, all included bodies, besides their translation from near the surrounding ones, partake likewise of their true motions; and though that translation were not made, they would not be really at rest, but only seem to be so. For the surrounding bodies stand in the like relation to the surrounded as the exterior part of a whole does to the interior, or as the shell does to the kernel; but if the shell moves, the kernel will also move, as being part of the whole, without any removal from near the shell.

A property, near akin to the preceding, is this, that if a place is moved, whatever is placed therein moves along with it; and therefore a body, which is moved from a place in motion, partakes also of the motion of its place. Upon which account, all motions, from places in motion, are no other than parts of entire and absolute motions; and every entire motion is composed of the motion of the body out of its first place, and the motion of this place out of its place; and so on, until we come to some immovable place, as in the before-mentioned example of the sailor. Wherefore, entire and absolute motions can be no otherwise determined than by immovable places; and for that reason I did before refer those absolute motions to immovable places, but relative ones to movable places. Now no other places are immovable but those that, from infinity to infinity, do all retain the same given position one to another; and upon this account must ever remain unmoved; and do thereby constitute immovable space.

The causes by which true and relative motions are distinguished, one from the other, are the forces impressed upon bodies to generate motion. True motion is neither generated nor altered, but by some force impressed upon the body moved; but relative motion may be generated or altered without any force impressed upon the body. For it is sufficient only to impress some force on other bodies with which the former is compared, that by their giving way, that relation may be changed, in which the relative rest or motion of this other body did consist. Again, true motion suffers always some change from any force impressed upon the moving body; but relative motion does not necessarily undergo any change by such forces. For if the same forces are likewise impressed on those other bodies, with which the comparison is made, that the relative position may be preserved, then that condition will be preserved in which the relative motion consists. And therefore any relative motion may be changed when the true motion remains unaltered, and the relative may be preserved when the true suffers some change. Thus, true motion by no means consists in such relations.

The effects which distinguish absolute from relative motion are, the forces of receding from the axis of circular motion. For there are no such forces in a circular motion purely relative, but in a true and absolute circular motion, they are greater or less, according to the quantity of the motion. If a vessel, hung by a long cord, is so often turned about that the cord is strongly twisted, then filled with water, and held at rest together with the water; thereupon, by the sudden action of another force, it is whirled about the contrary way, and while the cord is untwisting itself, the vessel continues for some time in this motion; the surface of the water will at first be plain, as before the vessel began to move; but after that, the vessel, by gradually communicating its motion to the water, will make it begin sensibly to revolve, and recede by little and little from the middle, and ascend to the sides of the vessel, forming itself into a concave figure (as I have experienced), and the swifter the motion becomes, the higher will the water rise, till at last, performing its revolutions in the same times with the vessel, it becomes relatively at rest in it. This ascent of the water shows its endeavor to recede from the axis of its motion; and the true and absolute circular motion of the water, which is here directly contrary to the relative, becomes known, and may be measured by this endeavor. At first, when the relative motion of the water in the vessel was greatest, it produced no endeavor to recede from the axis; the water showed no tendency to the circumference, nor any ascent towards the sides of the vessel, but remained of a plain surface, and therefore its true circular motion had not yet begun. But afterwards, when the relative motion of the water had decreased, the ascent thereof towards the sides of the vessel proved its endeavor to recede from the axis; and this endeavor showed the real circular motion of the water continually increasing, till it had acquired its greatest quantity, when the water rested relatively in the vessel. And therefore this endeavor does not depend upon any translation of the water in respect of the ambient bodies, nor can true circular motion be defined by such translation. There is only one real circular motion of any one revolving body, corresponding to only one power of endeavoring to recede from its axis of motion, as its proper and adequate effect; but relative motions, in one and the same body, are innumerable, according to the various relations it bears to external bodies, and, like other relations, are altogether destitute of any real effect, any otherwise than they may perhaps partake of that one only true motion. And therefore in their system who suppose that our heavens, revolving below the sphere of the fixed stars, carry the planets along with them; the several parts of those heavens, and the planets, which are indeed relatively at rest in their heavens, do yet really move. For they change their position one to another (which never happens to bodies truly at rest), and being carried together with their heavens, partake of their motions, and as parts of revolving wholes, endeavor to recede from the axis of their motions.

Wherefore relative quantities are not the quantities themselves, whose names they bear, but those sensible measures of them (either accurate or inaccurate), which are commonly used instead of the measured quantities themselves. And if the meaning of words is to be determined by their use, then by the names time, space, place, and motion, their [sensible] measures are properly to be understood; and the expression will be unusual, and purely mathematical, if the measured quantities themselves are meant. On this account, those violate the accuracy of language, which ought to be kept precise, who interpret these words for the measured quantities. Nor do those less defile the purity of mathematical and philosophical truths, who confound real quantities with their relations and sensible measures.

It is indeed a matter of great difficulty to discover, and effectually to distinguish, the true motions of particular bodies from the apparent; because the parts of that immovable space, in which those motions are performed, do by no means come under the observation of our senses. Yet the thing is not altogether desperate; for we have some arguments to guide us, partly from the apparent motions, which are the differences of the true motions; partly from the forces, which are the causes and effects of the true motions. For instance, if two globes, kept at a given distance one from the other by means of a cord that connects them, were revolved about their common centre of gravity, we might, from the tension of the cord, discover the endeavor of the globes to recede from the axis of their motion, and from thence we might compute the quantity of their circular motions. And then if any equal forces should be impressed at once on the alternate faces of the globes to augment or diminish their circular motions, from the increase or decrease of the tension of the cord, we might infer the increment or decrement of their motions; and thence would be found on what faces those forces ought to be impressed, that the motions of the globes might be most augmented; that is, we might discover their hindmost faces, or those which, in the circular motion, do follow. But the faces which follow being known, and consequently the opposite ones that precede, we should likewise know the determination of their motions. And thus we might find both the quantity and the determination of this circular motion, even in an immense vacuum, where there was nothing external or sensible with which the globes could be compared. But now, if in that space some remote bodies were placed that kept always a given position one to another, as the fixed stars do in our regions, we could not indeed determine from the relative translation of the globes among those bodies, whether the motion did belong to the globes or to the bodies. But if we observed the cord, and found that its tension was that very tension which the motions of the globes required, we might conclude the motion to be in the globes, and the bodies to be at rest; and then, lastly, from the translation of the globes among the bodies, we should find the determination of their motions. But how we are to obtain the true motions from their causes, effects, and apparent differences, and the converse, shall be explained more at large in the following treatise. For to this end it was that I composed it.

AXIOMS, OR LAWS OF MOTION

LAW I

Every body continues in its state of rest, or of uniform motion in a right line, unless it is compelled to change that state by forces impressed upon it.

Projectiles continue in their motions, so far as they are not retarded by the resistance of the air, or impelled downwards by the force of gravity. A top, whose parts by their cohesion are continually drawn aside from rectilinear motions, does not cease its rotation, otherwise than as it is retarded by the air. The greater bodies of the planets and comets, meeting with less resistance in freer spaces, preserve their motions both progressive and circular for a much longer time.

LAW II

The change of motion is proportional to the motive force impressed; and is made in the direction of the right line in which that force is impressed.

If any force generates a motion, a double force will generate double the motion, a triple force triple the motion, whether that force be impressed altogether and at once, or gradually and successively. And this motion (being always directed the same way with the generating force), if the body moved before, is added to or subtracted from the former motion, according as they directly conspire with or are directly contrary to each other; or obliquely joined, when they are oblique, so as to produce a new motion compounded from the determination of both.

LAW III

To every action there is always opposed an equal reaction: or, the mutual actions of two bodies upon each other are always equal, and directed to contrary parts.

Whatever draws or presses another is as much drawn or pressed by that other. If you press a stone with your finger, the finger is also pressed by the stone. If a horse draws a stone tied to a rope, the horse (if I may so say) will be equally drawn back towards the stone; for the distended rope, by the same endeavor to relax or unbend itself, will draw the horse as much towards the stone as it does the stone towards the horse, and will obstruct the progress of the one as much as it advances that of the other. If a body impinge upon another, and by its force change the motion of the other, that body also (because of the equality of the mutual pressure) will undergo an equal change, in its own motion, towards the contrary part. The changes made by these actions are equal, not in the velocities but in the motions of bodies; that is to say, if the bodies are not hindered by any other impediments. For, because the motions are equally changed, the changes of the velocities made towards contrary parts are inversely proportional to the bodies. This law takes place also in attractions, as will be proved in the next Scholium.

248

RULES OF REASONING IN PHILOSOPHY

RULE I

We are to admit no more causes of natural things than such as are both true and sufficient to explain their appearances.

To this purpose the philosophers say that Nature does nothing in vain, and more is in vain when less will serve; for Nature is pleased with simplicity, and affects not the pomp of superfluous causes.

RULE II

Therefore to the same natural effects we must, as far as possible, assign the same causes.

As to respiration in a man and in a beast; the descent of stones in Europe and in America; the light of our culinary fire and of the sun; the reflection of light in the earth, and in the planets.

RULE III

The qualities of bodies, which admit neither intensification nor remission of degrees, and which are found to belong to all bodies within the reach of our experiments, are to be esteemed the universal qualities of all bodies whatsoever.

For since the qualities of bodies are only known to us by experiments, we are to hold for universal all such as universally agree with experiments; and such as are not liable to diminution can never be quite taken away. We are certainly not to relinquish the evidence of experiments for the sake of dreams and vain fictions of our own devising; nor are we to recede from the analogy of Nature, which is wont to be simple, and always consonant to itself. We no other way know the extension of bodies than by our senses, nor do these reach it in all bodies; but because we perceive extension in all that are sensible, therefore we ascribe it universally to all others also. That abundance of bodies are hard, we learn by experience; and because the hardness of the whole arises from the hardness of the parts, we therefore justly infer the hardness of the undivided particles not only of the bodies we feel but of all others. That all bodies are impenetrable, we gather not from reason, but from sensation. The bodies which we handle we find impenetrable, and thence conclude impenetrability to be an universal property of all bodies whatsoever. That all bodies are movable, and endowed with certain powers (which we call the inertia) of persevering in their motion, or in their rest, we only infer from the like properties observed in the bodies which we have seen. The extension, hardness, impenetrability, mobility, and inertia of the whole, result from the extension, hardness, impenetrability, mobility, and inertia of the parts; and hence we conclude the least particles of all bodies to be also all extended, and hard and impenetrable, and movable, and endowed with their proper inertia. And this is the foundation of all philosophy. Moreover, that the divided but contiguous particles of bodies may be

249

separated from one another. is matter of observation; and, in the particles that remain undivided. our minds are able to distinguish yet lesser parts, as is mathematically demonstrated. But whether the parts so distinguished, and not yet divided. may. by the powers of Nature, be actually divided and separated from one another, we cannot certainly determine. Yet, had we the proof of but one experiment that any undivided particle, in breaking a hard and solid body, suffered a division. we might by virtue of this rule conclude that the undivided as well as the divided particles may be divided and actually separated to infinity.

Lastly, if it universally appears. by experiments and astronomical observations, that all bodies about the earth gravitate towards the earth, and that in proportion to the quantity of matter which they severally contain; that the moon likewise, according to the quantity of its matter, gravitates towards the earth; that, on the other hand. our sea gravitates towards the moon; and all the planets one towards another; and the comets in like manner towards the sun; we must, in consequence of this rule, universally allow that all bodies whatsoever are endowed with a principle of mutual gravitation. For the argument from the appearances concludes with more force for the universal gravitation of all bodies than for their impenetrability; of which, among those in the celestial regions. we have no experiments, nor any manner of observation. Not that I affirm gravity to be essential to bodies: by their *vis insita* I mean nothing but their inertia. This is immutable. Their gravity is diminished as they recede from the earth.

RULE IV

In experimental philosophy we are to look upon propositions inferred by general induction from phenomena as accurately or very nearly true, notwithstanding any contrary hypotheses that may be imagined, till such time as other phenomena occur, by which they may either be made more accurate, or liable to exceptions.

This rule we must follow, that the argument of induction may not be evaded by hypotheses.

GENERAL SCHOLIUM

THE hypothesis of vortices is pressed with many difficulties. That every planet by a radius drawn to the sun may describe areas proportional to the times of description, the periodic times of the several parts of the vortices should observe the square of their distances from the sun; but that the periodic times of the planets may obtain the ½th power of their distances from the sun, the periodic times of the parts of the vortex ought to be as the ³⁄₂th power of their distances. That the smaller vortices may maintain their lesser revolutions about Saturn, Jupiter, and other planets, and swim quietly and undisturbed in the greater vortex of the sun, the periodic times of the parts of the sun's vortex should be equal; but the rotation of the sun and planets about their axes, which ought to correspond with the motions of their vortices, recede far from all these proportions. The motions of the comets are exceedingly regular, are governed by the same laws with the motions of the planets, and can by no means be accounted for by the hypothesis of vortices; for comets are carried with very eccentric motions through all parts of the heavens indifferently, with a freedom that is incompatible with the notion of a vortex.

Bodies projected in our air suffer no resistance but from the air. Withdraw the air, as is done in Mr. Boyle's vacuum, and the resistance ceases; for in this void a bit of fine down and a piece of solid gold descend with equal velocity. And the same argument must apply to the celestial spaces above the earth's atmosphere; in these spaces, where there is no air to resist their motions, all bodies will move with the greatest freedom; and the planets and comets will constantly pursue their revolutions in orbits given in kind and position, according to the laws above explained; but though these bodies may, indeed, continue in their orbits by the mere laws of gravity, yet they could by no means have at first derived the regular position of the orbits themselves from those laws.

The six primary planets are revolved about the sun in circles concentric with the sun, and with motions directed towards the same parts, and almost in the same plane. Ten moons are revolved about the earth, Jupiter, and Saturn, in circles concentric with them, with the same direction of motion, and nearly in the planes of the orbits of those planets; but it is not to be conceived that mere mechanical causes could give birth to so many regular motions, since the comets range over all parts of the heavens in very eccentric orbits; for by that kind of motion they pass easily through the orbs of the planets, and with great rapidity; and in their aphelions, where they move the slowest, and are detained the longest, they recede to the greatest distances from each other, and hence suffer the least disturbance from their mutual attractions. This most beautiful system of the sun, planets, and comets, could only proceed from the counsel and dominion of an intelligent and powerful Being. And if the fixed stars are the centres of other like systems, these, being formed by the like wise counsel, must be all subject to the dominion of One; especially since the light of the fixed stars is of the same nature with the light of the sun, and from every system light passes into all the other systems: and lest the systems of the fixed stars should, by their gravity, fall on each other, he hath placed those systems at immense distances from one another.

This Being governs all things, not as the soul of the world, but as Lord over all; and on account of his dominion he is wont to be called *Lord God* παντοκρά-τωρ, or *Universal Ruler;* for *God* is a relative word, and has a respect to servants; and *Deity* is the dominion of God not over his own body, as those imagine who fancy God to be the soul of the world, but over servants. The Supreme God is a Being eternal, infinite, absolutely perfect; but a being, however perfect, without dominion, cannot be said to be Lord God; for we say, my God, your God, the God of Israel, the God of Gods, and Lord of Lords; but we do not say, my Eternal, your Eternal, the Eternal of Israel, the Eternal of Gods; we do not say, my Infinite, or my Perfect: these are titles which have no respect to servants. The word *God*[1] usually signifies *Lord;* but every lord is not a God. It is the dominion of a spiritual being which constitutes a God: a true, supreme, or imaginary dominion makes a true, supreme, or imaginary God. And from his true dominion it follows that the true God is a living, intelligent, and powerful Being; and, from his other perfections, that he is supreme, or most perfect. He is eternal and infinite, omnipotent and omniscient; that is, his duration reaches from eternity to eternity; his presence from infinity to infinity; he governs all things, and knows all things that are or can be done. He is not eternity and infinity, but eternal and infinite; he is not duration or space, but he endures and is present. He endures forever, and is everywhere present; and, by existing always and everywhere, he constitutes duration and space. Since every particle of space is *always*, and every indivisible moment of duration is *everywhere*, certainly the Maker and Lord of all things cannot be *never* and *nowhere*. Every soul that has perception is, though in different times and in different organs of sense and motion, still the same indivisible person. There are given successive parts in duration, coexistent parts in space, but neither the one nor the other in the person of a man, or his thinking principle; and much less can they be found in the thinking substance of God. Every man, so far as he is a thing that has perception, is one and the same man during his whole life, in all and each of his organs of sense. God is the same God, always and everywhere. He is omnipresent not *virtually* only, but also *substantially;* for virtue cannot subsist without substance. In him[2] are all things contained and moved; yet neither affects the other: God suffers nothing from the motion of bodies; bodies find no resistance from the omnipresence of God. It is allowed by all that the Supreme God exists necessarily; and by the same necessity he

[1]Dr. Pocock derives the Latin word *Deus* from the Arabic *du* (in the oblique case *di*), which signifies *Lord.* And in this sense princes are called *gods*, Psalms, 82.6; and John, 10.35. And Moses is called a *god* to his brother Aaron, and a *god* to Pharaoh, Exodus, 4.16; and 7.1. And in the same sense the souls of dead princes were formerly, by the heathens, called *gods*, but falsely, because of their want of dominion.

[2]This was the opinion of the ancients. So Pythagoras, in Cicero *De natura deorum* i. Thales, Anaxagoras, Virgil, in *Georgics* iv. 220; and *Aeneid* vi. 721. Philo, *Allegories*, at the beginning of Book I. Aratus, in his *Phænomena*, at the beginning. So also the sacred writers: as St. Paul, in Acts, 17.27, 28. St. John's Gospel, 14.2. Moses, in Deuteronomy, 4.39; and 10.14. David, in Psalms, 139.7,8,9. Solomon, in I Kings, 8.27. Job, 22.12,13,14. Jeremiah, 23.23,24. The idolaters supposed the sun, moon, and stars, the souls of men, and other parts of the world, to be parts of the Supreme God, and therefore to be worshipped; but erroneously.

6

Michael Faraday, THE CHEMICAL HISTORY OF A CANDLE

1. Compare and contrast the style of Faraday's writing to that of Galileo and Newton. Whom would you have preferred to have heard lecture?

2. Faraday argued that "There is no more open door by which you can enter the study of natural philosophy than by considering the physical phenomena of a candle." Why do you suppose he chose a candle as the vehicle for a discussion of physical science?

3. How many distinct physical phenomena (e.g. convection of heated air, melting of wax) can you identify as you read this selection?

     Michael Faraday (1791-1867) began his career as a laboratory assistant to Sir Humphrey Davy, the inventor of a safety lamp for miners and one of the foremost chemists of his day. Although he was self-taught, Faraday eventually became the Director of the Royal Institution and Professor of Chemistry. Regarded by many as the most gifted experimenter who ever lived, Faraday not only made wide-ranging discoveries in chemistry and electromagnetism, but also shared those discoveries with the public as one of the most popular lecturers of his day.

THE CHEMICAL HISTORY
OF A CANDLE

A COURSE OF LECTURES DELIVERED BEFORE A
JUVENILE AUDIENCE AT THE ROYAL
INSTITUTION

LECTURE I

A CANDLE: THE FLAME—ITS SOURCES—STRUCTURE—
MOBILITY—BRIGHTNESS

I PURPOSE, in return for the honor you do us by coming to see what are our proceedings here, to bring before you, in the course of these lectures, the Chemical History of a Candle. I have taken this subject on a former occasion, and, were it left to my own will, I should prefer to repeat it almost every year, so abundant is the interest that attaches itself to the subject, so wonderful are the varieties of outlet which it offers into the various departments of philosophy. There is not a law under which any part of this universe is governed which does not come into play and is touched upon in these phenomena. There is no better, there is no more open door by which you can enter into the study of natural philosophy than by considering the physical phenomena of a candle. I trust, therefore, I shall not disappoint you in choosing this for my subject rather than any newer topic, which could not be better, were it even so good.

And, before proceeding, let me say this also: that, though our subject be so great, and our intention that of treating it honestly, seriously, and philosophically, yet I mean to pass away from all those who are seniors among us. I claim the privilege of speaking to juveniles as a juvenile myself. I have done so on former occasions, and, if you please, I shall do so again. And, though I stand here with the knowledge of having the words I utter given to the world, yet that shall not deter me from speaking in the same familiar way to those whom I esteem nearest to me on this occasion.

And now, my boys and girls, I must first tell you of what candles are made. Some are great curiosities. I have here some bits of timber, branches of trees particularly famous for their burning. And here you see a piece of that very curious substance, taken out of some of the bogs in Ireland, called *candle-wood;* a hard, strong, excellent wood, evidently fitted for good work as a register of force, and yet, withal, burning so well that where it is found they make splinters of it, and torches, since it burns like a candle, and gives a very good light indeed. And in this wood we have one of the most beautiful illustrations of the general nature of a candle that I can possibly give. The fuel provided, the means of bringing that fuel to the place of chemical action, the regular and gradual supply of air to that place of action—heat and light—all produced by a little piece of wood of this kind, forming, in fact, a natural candle.

But we must speak of candles as they are in commerce. Here are a couple of candles commonly called dips. They are made of lengths of cotton cut off, hung up by a loop, dipped into melted tallow, taken out again and cooled, then redipped, until there is an accumulation of tallow round the cotton. In order that you may have an idea of the various characters of these candles, you see these which I hold in my hand—they are very small and very curious. They are, or were, the candles used by the miners in coal mines. In olden times the miner had to find his own candles, and it was supposed that a small candle would not so soon set fire to the fire-damp in the coal mines as a large one; and for that reason, as well as for economy's sake, he had candles made of this sort—20, 30, 40, or 60 to the pound. They have been replaced since then by the steel-mill, and then by the Davy lamp, and other safety lamps of various kinds. I have here a candle that was taken out of the *Royal George*([1]), it is said, by Colonel Pasley. It has been sunk in the sea for many years, subject to the action of salt water. It shows you how well candles may be preserved; for, though it is cracked about and broken a great deal,

[1] The *Royal George* sunk at Spithead on the 29th of August, 1782. Colonel Pasley commenced operations for the removal of the wreck by the explosion of gunpowder, in August, 1839. The candle which Professor Faraday exhibited must therefore have been exposed to the action of salt water for upward of fifty-seven years.

yet when lighted it goes on burning regularly, and the tallow resumes its natural condition as soon as it is fused.

Mr. Field, of Lambeth, has supplied me abundantly with beautiful illustrations of the candle and its materials; I shall therefore now refer to them. And, first, there is the suet—the fat of the ox— Russian tallow, I believe, employed in the manufacture of these dips, which Gay-Lussac, or some one who intrusted him with his knowledge, converted into that beautiful substance, stearin, which you see lying beside it. A candle, you know, is not now a greasy thing like an ordinary tallow candle, but a clean thing, and you may almost scrape off and pulverize the drops which fall from it without soiling any thing. This is the process he adopted([2]): The fat or tallow is first boiled with quick-lime, and made into a soap, and then the soap is decomposed by sulphuric acid, which takes away the lime, and leaves the fat rearranged as stearic acid, while a quantity of glycerin is produced at the same time. Glycerin—absolutely a sugar, or a substance similar to sugar—comes out of the tallow in this chemical change. The oil is then pressed out of it; and you see here this series of pressed cakes, showing how beautifully the impurities are carried out by the oily part as the pressure goes on increasing, and at last you have left that substance, which is melted, and cast into candles as here represented. The candle I have in my hand is a stearin candle, made of stearin from tallow in the way I have told you. Then here is a sperm candle, which comes from the purified oil of the spermaceti whale. Here, also, are yellow beeswax and refined beeswax, from which candles are made. Here, too, is that curious substance called paraffine, and some paraffine candles, made of paraffine obtained from the bogs of Ireland. I have here also a substance brought from Japan since we have forced an entrance into

[2] The fat or tallow consists of a chemical combination of fatty acids with glycerin. The lime unites with the palmitic, oleic, and stearic acids, and separates the glycerin. After washing, the insoluble lime soap is decomposed with hot dilute sulphuric acid. The melted fatty acids thus rise as an oil to the surface, when they are decanted. They are again washed and cast into thin plates, which, when cold, are placed between layers of cocoanut matting and submitted to intense hydraulic pressure. In this way the soft oleic acid is squeezed out, while the hard palmitic and stearic acids remain. These are farther purified by pressure at a higher temperature and washing in warm dilute sulphuric acid, when they are ready to be made into candles. These acids are harder and whiter than the fats from which they were obtained, while at the same time they are cleaner and more combustible.

that out-of-the-way place—a sort of wax which a kind friend has sent me, and which forms a new material for the manufacture of candles.

And how are these candles made? I have told you about dips, and I will show you how moulds are made. Let us imagine any of these candles to be made of materials which can be cast. "Cast!" you say. "Why, a candle is a thing that melts, and surely if you can melt it you can cast it." Not so. It is wonderful, in the progress of manufacture, and in the consideration of the means best fitted to produce the required result, how things turn up which one would not expect beforehand. Candles can not always be cast. A wax candle can never be cast. It is made by a particular process which I can illustrate in a minute or two, but I must not spend much time on it. Wax is a thing which, burning so well, and melting so easily in a candle, can not be cast. However, let us take a material that can be cast. Here is a frame, with a number of moulds fastened in it. The first thing to be done is to put a wick through them. Here is one—a plaited wick, which does not require snuffing([1])—supported by a little wire. It goes to the bottom, where it is pegged in—the little peg holding the cotton tight, and stopping the aperture so that nothing fluid shall run out. At the upper part there is a little bar placed across, which stretches the cotton and holds it in the mould. The tallow is then melted, and the moulds are filled. After a certain time, when the moulds are cool, the excess of tallow is poured off at one corner, and then cleaned off altogether, and the ends of the wick cut away. The candles alone then remain in the mould, and you have only to upset them, as I am doing, when out they tumble, for the candles are made in the form of cones, being narrower at the top than at the bottom; so that, what with their form and their own shrinking, they only need a little shaking, and out they fall. In the same way are made these candles of stearin and of paraffine. It is a curious thing to see how wax candles are made. A lot of cottons are hung upon frames, as you see here, and covered with metal tags at the ends to keep the wax from covering the cotton in those places. These are carried to a heater, where the wax is melted.

[1] A little borax or phosphorus salt is sometimes added in order to make the ash fusible.

As you see, the frames can turn round; and, as they turn, a man takes a vessel of wax and pours it first down one, and then the next, and the next, and so on. When he has gone once round. if it is sufficiently cool, he gives the first a second coat, and so on until they are all of the required thickness. When they have been thus clothed, or fed, or made up to that thickness, they are taken off and placed elsewhere. I have here, by the kindness of Mr. Field, several specimens of these candles. Here is one only half finished. They are then taken down and well rolled upon a fine stone slab, and the conical top is moulded by properly shaped tubes, and the bottoms cut off and trimmed. This is done so beautifully that they can make candles in this way weighing exactly four or six to the pound, or any number they please.

We must not, however, take up more time about the mere manufacture, but go a little farther into the matter. I have not yet referred you to luxuries in candles (for there is such a thing as luxury in candles). See how beautifully these are colored; you see here mauve, magenta, and all the chemical colors recently introduced, applied to candles. You observe, also, different forms employed. Here is a fluted pillar most beautifully shaped; and I have also here some candles sent me by Mr. Pearsall, which are ornamented with designs upon them, so that, as they burn, you have, as it were, a glowing sun above, and bouquet of flowers beneath. All, however, that is fine and beautiful is not useful. These fluted candles, pretty as they are, are bad candles; they are bad because of their external shape. Nevertheless, I show you these specimens, sent to me from kind friends on all sides, that you may see what is done and what may be done in this or that direction; although, as I have said, when we come to these refinements, we are obliged to sacrifice a little in utility.

Now as to the light of the candle. We will light one or two, and set them at work in the performance of their proper functions. You observe a candle is a very different thing from a lamp. With a lamp you take a little oil, fill your vessel, put in a little moss or some cotton prepared by artificial means, and then light the top of the wick. When the flame runs down the cotton to the oil, it gets extinguished, but it goes on burning in the part above. Now I have no doubt you will ask how it is that the oil which will not burn of itself gets up to the top of the cotton, where it will burn. We shall presently examine that; but there is a much more wonderful thing about the burning of a candle than this. You have here a solid substance with no vessel to contain it; and how is it that this solid substance can get up to the place where the flame is? How is it that this solid gets there, it not being a fluid? or, when it is made a fluid, then how is it that it keeps together? This is a wonderful thing about a candle.

We have here a good deal of wind, which will help us in some of our illustrations, but tease us in others; for the sake, therefore, of a little regularity, and to simplify the matter, I shall make a quiet flame, for who can study a subject when there are difficulties in the way not belonging to it? Here is a clever invention of some costermonger or street-stander in the market-place for the shading of their candles on Saturday nights, when they are selling their greens, or potatoes, or fish. I have very often admired it. They put a lamp-glass round the candle, supported on a kind of gallery, which clasps it, and it can be slipped up and down as required. By the use of this lamp-glass, employed in the same way, you have a steady flame, which you can look at, and carefully examine, as I hope you will do, at home.

You see, then, in the first instance, that a beautiful cup is formed. As the air comes to the candle, it moves upward by the force of the current which the heat of the candle produces, and it so cools all the sides of the wax, tallow, or fuel as to keep the edge much cooler than the part within; the part within melts by the flame that runs down the wick as far as it can go before it is extinguished, but the part on the outside does not melt. If I made a current in one direction, my cup would be lop-sided, and the fluid would consequently run óver; for the same force of gravity which holds worlds together holds this fluid in a horizontal position, and if the cup be not horizontal, of course the fluid will run away in guttering. You see, therefore, that the cup is formed by this beautifully regular ascending current of air playing upon all sides, which keeps the exterior of the candle cool. No fuel would serve for a candle which has not the property of giving this cup, except such fuel as the Irish bogwood, where the material itself is like a sponge and holds its own fuel. You see now why you would have had such a bad result if you were to burn these beautiful candles that I have shown you, which are irregular, intermittent in their shape, and can not, therefore, have that nicely-formed edge to the cup which is the great beauty in a candle. I hope you will now see that the perfection of a process—that is, its utility—is the better point of beauty about it. It is not the best looking thing, but the best acting thing, which is the most advantageous to us. This good-looking candle is a bad-burning one. There will be a guttering round about it because of the irregularity of the stream of air and the badness of the cup which is formed thereby. You may see some pretty examples (and I trust you will notice these instances) of the action of the ascending current when you have a little gutter run down the side of a candle, making it thicker there than it is elsewhere. As the candle goes on burning, that keeps its place and forms a little pillar sticking up by the side, because, as it rises higher above the rest of the wax or fuel, the air gets better round it, and it is more cooled and better able to resist the action of the heat at a little distance. Now the greatest mistakes and faults with regard to candles, as in many other things, often bring with them instruction which we should not receive if they had not occurred. We come here to be philosophers, and I hope you will always remember that whenever a result happens, especially if it be new, you should say, "What is the cause? Why does it occur?" and you will, in the course of time, find out the reason.

Then there is another point about these candles which will answer a question—that is, as to the way in which this fluid gets out of the cup, up the wick, and into the place of combustion. You know that the flames on these burning wicks in candles made of beeswax, stearin, or spermaceti, do not run down to the wax or other matter, and melt it all away, but keep to their own right place. They are fenced off from the fluid below, and do not encroach on the cup at the sides. I can not imagine a more beautiful example than the condition of adjustment under which a candle makes one part subserve to the other to the very end of its action. A combustible thing like that, burning away gradually, never being intruded upon by the flame, is a very beautiful sight, especially when you come to learn what a vigorous thing flame is—what power it has of destroying the wax itself when it gets hold of it, and of disturbing its proper form if it come only too near.

But how does the flame get hold of the fuel? There is a beautiful point about that—*capillary attraction.*([4]) "Capillary attraction!" you say—"the attraction of hairs." Well, never mind the name; it was given in old times, before we had a good understanding of what the real power was. It is by what is called capillary attraction that the fuel is conveyed to the part where combustion goes on, and is deposited there, not in a careless way, but very beautifully in the very midst of the centre of action, which takes place around it. Now I am going to give you one or two instances of capillary attraction. It is that kind of action or attraction which makes two things that do not dissolve in each other still hold together. When you wash your hands, you wet them thoroughly; you take a little soap to make the adhesion better, and you find your hands remain wet. This is by that kind of attraction of which I am about to speak. And, what is more, if your hands are not soiled (as they almost always are by the usages of life), if you put your finger into a little warm water, the water will creep a little way up the finger, though you may not stop to examine it. I have here a substance which is rather porous—a column

of salt—and I will pour into the plate at the bottom, not water, as it appears, but a saturated solution of salt which can not absorb more, so that the action which you see will not be due to its dissolving any thing. We may consider the plate to be the candle, and the salt the wick, and this solution the melted tallow. (I have colored the fluid, that you may see the action better.) You observe that, now

FIG. 55

I pour in the fluid, it rises and gradually creeps up the salt higher and higher (FIG. 55); and provided the column does not tumble over, it will go to the top. If this blue solution were combustible, and we were to place a wick at the top of the salt, it would burn as it entered into

[4] Capillary attraction or repulsion is the cause which determines the ascent or descent of a fluid in a capillary tube. If a piece of thermometer tubing, open at each end, be plunged into water, the latter will instantly rise in the tube considerably above its external level. If, on the other hand, the tube be plunged into mercury, a repulsion instead of attraction will be exhibited, and the level of the mercury will be lower in the tube than it is outside.

the wick. It is a most curious thing to see this kind of action taking place, and to observe how singular some of the circumstances are about it. When you wash your hands, you take a towel to wipe off the water; and it is by that kind of wetting, or that kind of attraction which makes the towel become wet with water, that the wick is made wet with the tallow. I have known some careless boys and girls (indeed, I have known it happen to careful people as well) who, having washed their hands and wiped them with a towel, have thrown the towel over the side of the basin, and before long it has drawn all the water out of the basin and conveyed it to the floor, because it happened to be thrown over the side in such a way as to serve the purpose of a siphon.(⁵) That you may the better see the way in which the substances act one upon another, I have here a vessel made of wire gauze filled with water, and you may compare it in its action to the cotton in one respect, or to a piece of calico in the other. In fact, wicks are sometimes made of a kind of wire gauze. You will observe that this vessel is a porous thing; for if I pour a little water on to the top, it will run out at the bottom. You would be puzzled for a good while if I asked you what the state of this vessel is, what is inside it, and why it is there? The vessel is full of water, and yet you see the water goes in and runs out as if it were empty. In order to prove this to you, I have only to empty it. The reason is this: the wire, being once wetted, remains wet; the meshes are so small that the fluid is attracted so strongly from the one side to the other, as to remain in the vessel, although it is porous. In like manner, the particles of melted tallow ascend the cotton and get to the top; other particles then follow by their mutual attraction for each other, and as they reach the flame they are gradually burned.

Here is another application of the same principle. You see this bit of cane. I have seen boys about the streets, who are very anxious to appear like men, take a piece of cane, and light it, and smoke it, as an imitation of a cigar. They are enabled to do so by the permeability

⁵ The late Duke of Sussex was, we believe, the first to show that a prawn might be washed upon this principle. If the tail, after pulling off the fan part, be placed in a tumbler of water, and the head be allowed to hang over the outside, the water will be sucked up the tail by capillary attraction, and will continue to run out through the head until the water in the glass has sunk so low that the tail ceases to dip into it.

of the cane in one direction, and by its capillarity. If I place this piece of cane on a plate containing some camphene (which is very much like paraffine in its general character), exactly in the same manner as the blue fluid rose through the salt will this fluid rise through the piece of cane. There being no pores at the side, the fluid can not go in that direction, but must pass through its length. Already the fluid is at the top of the cane; now I can light it and make it serve as a candle. The fluid has risen by the capillary attraction of the piece of cane, just as it does through the cotton in the candle.

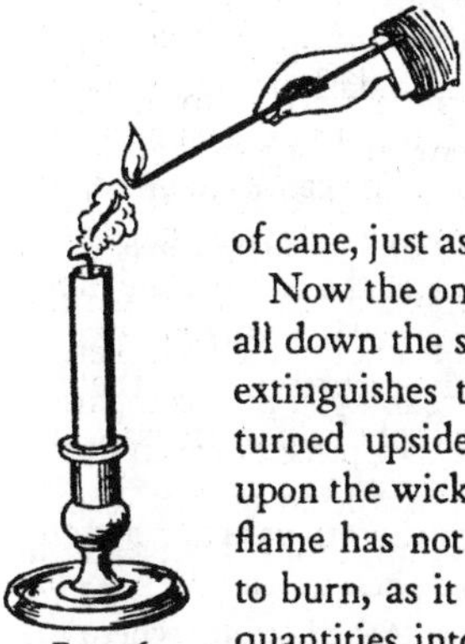

FIG. 56

Now the only reason why the candle does not burn all down the side of the wick is that the melted tallow extinguishes the flame. You know that a candle, if turned upside down, so as to allow the fuel to run upon the wick, will be put out. The reason is, that the flame has not had time to make the fuel hot enough to burn, as it does above, where it is carried in small quantities into the wick, and has all the effect of the heat exercised upon it.

There is another condition which you must learn as regards the candle, without which you would not be able fully to understand the philosophy of it, and that is the vaporous condition of the fuel. In order that you may understand that, let me show you a very pretty but very commonplace experiment. If you blow a candle out cleverly, you will see the vapor rise from it. You have, I know, often smelt the vapor of a blown-out candle, and a very bad smell it is; but if you blow it out cleverly you will be able to see pretty well the vapor into which this solid matter is transformed. I will blow out one of these candles in such a way as not to disturb the air around it by the continuing action of my breath; and now, if I hold a lighted taper two or three inches from the wick, you will observe a train of fire going through the air till it reaches the candle (FIG. 56). I am obliged to be quick and ready, because if I allow the vapor time to cool, it becomes condensed into a liquid or solid, or the stream of combustible matter gets disturbed.

Now as to the shape or form of the flame. It concerns us much to

263

know about the condition which the matter of the candle finally assumes at the top of the wick, where you have such beauty and brightness as nothing but combustion or flame can produce. You have the glittering beauty of gold and silver, and the still higher lustre of jewels like the ruby and diamond; but none of these rival the brilliancy and beauty of flame. What diamond can shine like flame? It owes its lustre at nighttime to the very flame shining upon it. The flame shines in darkness, but the light which the diamond has is as nothing until the flame shines upon it, when it is brilliant again. The candle alone shines by itself and for itself, or for those who have arranged the materials. Now let us look a little at the form of the flame as you see it under the glass shade. It is steady and equal, and its general form is that which is represented in the diagram (Fig. 57), varying with atmospheric disturbances, and also varying according to the size of the candle. It is a bright oblong, brighter at the top than toward the bottom, with the wick in the middle, and, besides the wick in the middle, certain darker parts towards the bottom, where the ignition is not so perfect as in the part above. I have a drawing here, sketched many years ago by Hooker, when he made his investigations. It is the drawing of the flame of a lamp, but it will apply to the flame of a candle. The cup of the candle is the vessel or lamp; the melted spermaceti is the oil; and the wick is common to both. Upon that he sets this little flame, and then he represents what is true, a certain quantity of matter rising about it which you do not see, and which, if you have not been here before, or are not familiar with the subject, you will not know of. He has here represented the parts of the surrounding atmosphere that are very essential to the flame, and that are always present with it. There is a current formed, which draws the flame out; for the flame which you see is really drawn out by the current, and drawn upward to a great height, just as Hooker has here shown you by that prolongation of the current in the diagram. You may see this by taking a lighted candle, and putting it in the sun so as to get its shadow thrown on a piece of paper. How remarkable it is that that thing which is light enough to produce shadows of other objects can be made to throw its own shadow on a piece

Fig. 57

of white paper or card, so that you can actually see streaming round the flame something which is not part of the flame, but is ascending and drawing the flame upward. Now I am going to imitate the sunlight by applying the voltaic battery to the electric lamp. You now see our sun and its great luminosity; and by placing a candle between it and the screen, we get the shadow of the flame. You observe the shadow of the candle and of the wick; then there is a darkish part, as represented in the diagram (Fig. 58), and then a part which is more distinct. Curiously enough, however, what we see in the shadow as the darkest part of the flame is, in reality, the brightest part; and here you see streaming upward the ascending current of hot air, as shown by Hooker, which draws out the flame, supplies it with air, and cools the sides of the cup of melted fuel.

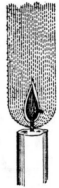

FIG. 58

I can give you here a little farther illustration, for the purpose of showing you how flame goes up or down according to the current. I have here a flame—it is not a candle flame—but you can, no doubt, by this time generalize enough to be able to compare one thing with another. What I am about to do is to change the ascending current that takes the flame upward into a descending current. This I can easily do by the little apparatus you see before me. The flame, as I have said, is not a candle flame, but it is produced by alcohol, so that it shall not smoke too much. I will also color the flame with another substance([6]), so that you may trace its course; for, with the spirit alone, you could hardly see well enough to have the opportunity of tracing its direction. By lighting this spirit of wine we have then a flame produced, and you observe that when held in the air it naturally goes upward. You understand now, easily enough, why flames go up under ordinary circumstances: it is because of the draught of air by which the combustion is formed. But now, by blowing the flame down, you see I am enabled to make it go downward into this little chimney, the direction of the current being changed (Fig. 59). Before we have concluded this course of lectures

[6] The alcohol had chloride of copper dissolved in it: this produces a beautiful green flame.

we shall show you a lamp in which the flame goes up and the smoke goes down, or the flame goes down and the smoke goes up. You see, then, that we have the power in this way of varying the flame in different directions.

There are now some other points that I must bring before you. Many of the flames you see here vary very much in their shape by  the currents of air blowing around them in different directions; but we can, if we like, make flames so that they will look like fixtures, and we can photograph them —indeed, we have to photograph them— so that they become fixed to us, if we wish to find out every thing concerning them. That, however, is not the only thing I wish to mention. If I take a flame sufficiently large, it does not keep that homogeneous, that uniform condition of shape, but it breaks out with a power of life which is quite wonderful. I am about to use another kind of fuel, but one which is truly and fairly a representative of the wax or tallow of a candle. I have here a large ball of cotton, which will serve as a wick. And, now that I have immersed it in spirit and applied a light to it, in what way does it differ from an ordinary candle? Why, it differs very much in one respect, that we have a vivacity and power about it, a beauty and a life entirely different from the light presented by a candle. You see those fine tongues of flame rising up. You have the same general disposition of the mass of the flame from below upward; but, in addition to that, you have this remarkable breaking out into tongues which you do not perceive in the case of a candle. Now, why is this? I must explain it to you, because, when you understand that perfectly, you will be able to follow me better in what I have to say hereafter. I suppose some here will have made for themselves the experiment I am going to show you. Am I right in supposing that any body here has played at snapdragon? I do not know a more beautiful illustration of the philosophy of flame, as to a certain part of its history, than the game of snapdragon. First, here is the dish; and let me say, that

FIG. 59

266

when you play snapdragon properly you ought to have the dish well warmed; you ought also to have warm plums, and warm brandy, which, however, I have not got. When you have put the spirit into the dish, you have the cup and the fuel; and are not the raisins acting like the wicks? I now throw the plums into the dish, and light the spirit, and you see those beautiful tongues of flame that I refer to. You have the air creeping in over the edge of the dish forming these

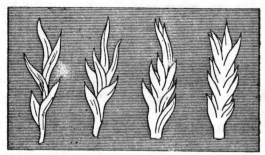

FIG. 60

tongues. Why? Because, through the force of the current and the irregularity of the action of the flame, it can not flow in one uniform stream. The air flows in so irregularly that you have what would otherwise be a single image broken up into a variety of forms, and each of these little tongues has an independent existence of its own. Indeed, I might say, you have here a multitude of independent candles. You must not imagine, because you see these tongues all at once, that the flame is of this particular shape. A flame of that shape is never so at any one time. Never is a body of flame, like that which you just saw rising from the ball, of the shape it appears to you. It consists of a multitude of different shapes, succeeding each other so fast that the eye is only able to take cognizance of them all at once. In former times I purposely analyzed a flame of that general character, and the diagram shows you the different parts of which it is composed (FIG. 60). They do not occur all at once; it is only because we see these shapes in such rapid succession that they seem to us to exist all at one time.

It is too bad that we have not got farther than my game of snap-dragon; but we must not, under any circumstances, keep you beyond your time. It will be a lesson to me in future to hold you more strictly to the philosophy of the thing than to take up your time so much with these illustrations.

LECTURE II

A CANDLE: BRIGHTNESS OF THE FLAME—AIR NECESSARY FOR COMBUSTION—PRODUCTION OF WATER

WE were occupied the last time we met in considering the general character and arrangement as regards the fluid portion of a candle, and the way in which that fluid got into the place of combustion. You see, when we have a candle burning fairly in a regular, steady atmosphere, it will have a shape something like the one shown in the diagram, and will look pretty uniform, although very curious in its character. And now I have to ask your attention to the means by which we are enabled to ascertain what happens in any particular part of the flame; why it happens; what it does in happening; and where, after all, the whole candle goes to; because, as you know very well, a candle being brought before us and burned, disappears, if burned properly, without the least trace of dirt in the candle stick; and this is a very curious circumstance. In order, then, to examine this candle carefully, I have arranged certain apparatus, the use of which you will see as I go on. Here is a candle; I am about to put the end of this glass tube into the middle of the flame—into that part which old Hooker has represented in the diagram as being rather dark, and which you can see at any time if you will look at a candle carefully, without blowing it about. We will examine this dark part first.

Now I take this bent glass tube, and introduce one end into that part of the flame, and you see at once that something is coming from the flame, out at the other end of the tube; and if I put a flask there, and leave it for a little while, you will see that something from the middle part of the flame is gradually drawn out, and goes through the tube, and into that flask, and there behaves very differently from what it does in the open air. It not only escapes from the end of the tube, but falls down to the bottom of the flask like a heavy substance, as indeed it is (Fig. 61). We find that this is the wax of the candle made into a vaporous fluid—not a gas. (You must learn the difference between a gas and a vapor: a gas remains permanent; a vapor is something that will condense.) If you blow out a candle, you perceive a very nasty smell, resulting from the condensation of this vapor. That is very different from what you have outside the flame; and, in order to make that more clear to you, I am about to produce and set fire to a larger portion of this vapor; for

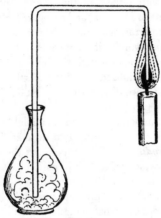

Fig. 61

what we have in the small way in a candle, to understand thoroughly, we must, as philosophers, produce in a larger way, if needful, that we may examine the different parts. And now Mr. Anderson will give me a source of heat, and I am about to show you what that vapor is. Here is some wax in a glass flask, and I am going to make it hot, as the inside of that candle flame is hot, and the matter about the wick is hot. [The lecturer placed some wax in a glass flask, and heated it over a lamp.] Now I dare say that is hot enough for me. You see that the wax I put in it has become fluid, and there is a little smoke coming from it. We shall very soon have the vapor rising up. I will make it still hotter, and now we get more of it, so that I can actually pour the vapor out of the flask into that basin, and set it on fire there. This, then, is exactly the same kind of vapor as we have in the middle of the candle; and that you may be sure this is the case, let us try whether we have not got here, in this flask, a real combustible vapor out of the middle of the candle. [Taking the flask into which the tube from the candle proceeded, and introducing a lighted taper (Fig. 62).] See how it burns. Now this is the vapor from the middle of the candle, produced by its own heat; and that is one of the first things you have to consider with respect to the progress of the wax in the course of its combustion, and as regards the changes it undergoes. I will arrange another tube carefully in the flame, and I should not wonder if we were able, by a little care, to get that vapor to pass through the tube to the other extremity, where we will light it, and obtain absolutely the flame of the candle at a place distant from it. Now, look at that. Is not that a very pretty experiment? Talk about laying on gas— why, we can actually lay on a candle! And you see from Fig. 62 this that there are clearly two different kinds of action— one the *production* of the vapor, and the other the *combustion* of it—both of which take place in particular parts of the candle.

I shall get no vapor from that part which is already burnt. If I raise the tube (Fig. 61) to the upper part of the flame, so soon as the vapor has been swept out what comes away will be no longer combustible; it is already burned. How burned? Why, burned thus: In the middle of the flame, where the wick is, there is this combustible vapor; on the outside of the flame is the air which we shall find necessary for the burning of the candle; between the two, intense chemical action takes place, whereby the air and the fuel act upon each other, and at the very same time that we obtain light the vapor inside is destroyed. If you examine where the heat of a candle is, you will find it very curiously arranged. Suppose I take this candle, and hold a piece of paper close upon the flame, where is the heat of that flame? Do you not see that it is *not* in the inside? It is in a ring, exactly in the place where I told you the chemical action was; and even in my irregular mode of making the experiment, if there is not too much disturbance, there will always be a ring. This is a good experiment for you to make at home. Take a strip of paper, have the air in the room quiet, and put the piece of paper right across the middle of the flame—(I must not talk while I make the experiment)—and you will find that it is burnt in two places, and that it is not burnt, or very little so, in the middle; and when you have tried the experiment once or twice, so as to make it nicely, you will be very interested to see where the heat is, and to find that it is where the air and the fuel come together.

This is most important for us as we proceed with our subject. Air is absolutely necessary for combustion; and, what is more, I must have you understand that *fresh* air is necessary, or else we should be imperfect in our reasoning and our experiments. Here is a jar of air; I place it over a candle, and it burns very nicely in it at first, showing that what I have said about it is true; but there will soon be a change. See how the flame is drawing upward, presently fading, and at last going out. And going out, why? Not because it wants air merely, for the jar is as full now as it was before; but it wants pure, fresh air. The jar is full of air, partly changed, partly not changed; but it does not contain sufficient of the fresh air which is necessary for the combustion of a candle. These are all points which we, as young chemists, have to gather up; and if we look a little more closely into this kind of action, we shall find certain steps of reasoning extremely interesting. For instance, here is the oil-lamp I showed you—an excellent lamp for our experiments—the old Argand lamp. I now make it like a candle [obstructing the passage of air into the centre of the flame]; there is the cotton; there is the oil rising up in it, and there is the conical flame. It burns poorly because there is a partial restraint of air. I have allowed no air to get to it save around the outside of the flame, and it does not burn well. I can not admit more air from the outside, because the wick is large; but if, as Argand did so cleverly, I open a passage to the middle of the flame, and so let air come in there, you will see how much more beautifully it burns. If I shut the air off, look how it smokes; and why? We have now some very interesting points to study: we have the case of the combustion of a candle; we have the case of a candle being put out by the want of air; and we have now the case of imperfect combustion, and this is to us so interesting that I want you to understand it as thoroughly as you do the case of a candle burning in its best possible manner. I will now make a great flame, because we need the largest possible illustrations. Here is a larger wick [burning turpentine on a ball of cotton]. All these things are the same as candles, after all. If we have larger wicks, we must have a larger supply of air, or we shall have less perfect combustion. Look, now, at this black substance going up into the atmosphere; there is a regular stream of it. I have provided means to carry off the imperfectly burned part, lest it should annoy you. Look at the soots that fly off from the flame; see what an imperfect combustion it is, because it can not get enough air. What, then, is happening? Why, certain things which are necessary to the combustion of a candle are absent, and very bad results are accordingly produced; but we see what happens to a candle when it is burnt in a pure and proper state of air. At the time when I showed you this charring by the ring of flame on the one side of the paper, I might have also shown you, by turning to the other side, that the burning of a candle produces the same kind of soot—charcoal, or carbon.

But, before I show that, let me explain to you, as it is quite necessary for our purpose, that, though I take a candle, and give you, as the general result, its combustion in the form of a flame, we must see whether combustion is always in this condition, or whether there are other conditions of flame; and we shall soon discover that there are, and that they are most important to us. I think, perhaps, the best illustration of such a point to us, as juveniles, is to show the result of strong contrast. Here is a little gunpowder. You know that gunpowder burns with flame; we may fairly call it flame. It contains carbon and other materials, which altogether cause it to burn with a flame. And here is some pulverized iron, or iron filings. Now I purpose burning these two things together. I have a little mortar in which I will mix them. (Before I go into these experiments, let me hope that none of you, by trying to repeat them for fun's sake, will do any harm. These things may all be very properly used if you take care, but without that much mischief will be done.) Well, then, here is a little gunpowder, which I put at the bottom of that little wooden vessel, and mix the iron filings up with it, my object being to make the gunpowder set fire to the filings and burn them in the air, and thereby show the difference between substances burning with flame and not with flame. Here is the mixture; and when I set fire to it you must watch the combustion, and you will see that it is of two kinds. You will see the gunpowder burning with a flame and the filings thrown up. You will see them burning, too, but without the production of flame. They will each burn separately. [The lecturer then ignited the mixture.] There is the gunpowder, which burns with a flame, and there are the filings: they burn with a different kind of combustion. You see, then, these two great distinctions; and upon these differences depend all the utility and all the beauty of flame which we use for the purpose of giving out light. When we use oil, or gas, or candle for the purpose of illumination, their fitness all depends upon these different kinds of combustion.

There are such curious conditions of flame that it requires some cleverness and nicety of discrimination to distinguish the kinds of combustion one from another. For instance, here is a powder which is very combustible, consisting, as you see, of separate little particles. It is called *lycopodium*,([7]) and each of these particles can produce a vapor, and produce its own flame; but, to see them burning, you would imagine it was all one flame. I will now set fire to a quantity, and you will see the effect. We saw a cloud of flame, apparently in one body; but that rushing noise [referring to the sound produced by the burning] was a proof that the combustion was not a continuous or regular one. This is the lightning of the pantomimes, and a very good imitation. [The experiment was twice repeated by blowing lycopodium from a glass tube through a spirit flame.] This is not an example of combustion like that of the filings I have been speaking of, to which we must now return.

Suppose I take a candle and examine that part of it which appears brightest to our eyes. Why, there I get these black particles, which already you have seen many times evolved from the flame, and which I am now about to evolve in a different way. I will take this candle and clear away the gutterage, which occurs by reason of the currents of air; and if I now arrange the glass tube so as just to dip into this luminous part, as in our first experiment, only higher, you see the

[7] Lycopodium is a yellowish powder found in the fruit of the club moss (*Lycopodium clavatum*). It is used in fireworks.

result. In place of having the same white vapor that you had before, you will now have a black vapor. There it goes, as black as ink. It is certainly very different from the white vapor; and when we put a light to it we shall find that it does not burn, but that it puts the light out. Well, these particles, as I said before, are just the smoke of the candle; and this brings to mind that old employment which Dean Swift recommended to servants for their amusement, namely, writing on the ceiling of a room with a candle. But what is that black substance? Why, it is the same carbon which exists in the candle. How comes it out of the candle? It evidently existed in the candle, or else we should not have had it here. And now I want you to follow me in this explanation. You would hardly think that all those substances which fly about London, in the form of soots and blacks, are the very beauty and life of the flame, and which are burned in it as those iron filings were burned here. Here is a piece of wire gauze, which will not let the flame go through it; and I think you will see, almost immediately, that when I bring it low enough to touch that part of the flame which is otherwise so bright, it quells and quenches it at once, and allows a volume of smoke to rise up.

I want you now to follow me in this point—that whenever a substance burns, as the iron filings burnt in the flame of gunpowder, without assuming the vaporous state (whether it becomes liquid or remains solid), it becomes exceedingly luminous. I have here taken three or four examples apart from the candle on purpose to illustrate this point to you, because what I have to say is applicable to all substances, whether they burn or whether they do not burn—that they are exceedingly bright if they retain their solid state, and that it is to this presence of solid particles in the candle flame that it owes its brilliancy.

Here is a platinum wire, a body which does not change by heat. If I heat it in this flame, see how exceedingly luminous it becomes. I will make the flame dim for the purpose of giving a little light only, and yet you will see that the heat which it can give to that platinum wire, though far less than the heat it has itself, is able to raise the platinum wire to a far higher state of effulgence. This flame has carbon in it; but I will take one that has no carbon in it. There is a material, a kind of fuel—a vapor, or gas, whichever you like to call it—in that vessel, and it has no solid particles in it; so I take that because it is an example of flame itself burning without any solid matter whatever; and if I now put this solid substance in it, you see what an intense heat it has, and how brightly it causes the solid body to glow. This is the pipe through which we convey this particular gas, which we call hydrogen, and which you shall know all about the next time we meet. And here is a substance called oxygen, by means of which this hydrogen can burn; and although we produce, by their mixture, far greater heat ([8]) than you can obtain from the candle, yet there is very little light. If, however, I take a solid substance, and put that into it, we produce an intense light. If I take a piece of lime, a substance which will not burn, and which will not vaporize by the heat (and because it does not vaporize remains solid, and remains heated), you will soon observe what happens as to its glowing. I have here a most intense heat produced by the burning of hydrogen in contact with the oxygen; but there is as yet very little light—not for want of heat, but for want of particles which can retain their solid state; but when I hold this piece of lime in the flame of the hydrogen as it burns in the oxygen, see how it glows! This is the glorious lime light, which rivals the voltaic light, and which is almost equal to sunlight. I have here a piece of carbon or charcoal, which will burn and give us light exactly in the same manner as if it were burnt as part of a candle. The heat that is in the flame of a candle decomposes the vapor of the wax, and sets free the carbon particles; they rise up heated and glowing as this now glows, and then enter into the air. But the particles, when burnt, never pass off from a candle in the form of carbon. They go off into the air as a perfectly invisible substance, about which we shall know hereafter.

Is it not beautiful to think that such a process is going on, and that such a dirty thing as charcoal can become so incandescent? You see it comes to this—that all bright flames contain these solid particles; all things that burn and produce solid particles, either during the time they are burning, as in the candle, or immediately after

[8] Bunsen has calculated that the temperature of the oxyhydrogen blowpipe is 8061° Centigrade. Hydrogen burning in air has a temperature of 3259° C., and coal gas in air, 2350° C.

being burnt, as in the case of the gunpowder and iron filings—all these things give us this glorious and beautiful light.

I will give you a few illustrations. Here is a piece of phosphorus, which burns with a bright flame. Very well; we may now conclude that phosphorus will produce, either at the moment that it is burning or afterwards, these solid particles. Here is the phosphorus lighted, and I cover it over with this glass for the purpose of keeping in what is produced. What is all that smoke? (FIG. 63.) That smoke consists of those very particles which are produced by the combustion of the phosphorus. Here, again, are two substances. This is chlorate of potassa, and this other sulphuret of antimony. I shall mix these together a little, and then they may

Fig. 63

be burnt in many ways. I shall touch them with a drop of sulphuric acid, for the purpose of giving you an illustration of chemical action, and they will instantly burn.([9]) [The lecturer then ignited the mixture by means of sulphuric acid.] Now, from the appearance of things, you can judge for yourselves whether they produce solid matter in burning. I have given you the train of reasoning which will enable you to say whether they do or do not; for what is this bright flame but the solid particles passing off?

Mr. Anderson has in the furnace a very hot crucible. I am about to throw into it some zinc filings, and they will burn with a flame like gunpowder. I make this experiment because you can make it well at home. Now I want you to see what will be the result of the combustion of this zinc. Here it is burning—burning beautifully like a candle, I may say. But what is all that smoke, and what are those little clouds of wool which will come to you if you can not come to them, and make themselves sensible to you in the form of the old philosophic wool, as it was called? We shall have left in that crucible, also, a quantity of this woolly matter. But I will take a piece of this

[9] The following is the action of the sulphuric in inflaming the mixture of sulphuret of antimony and chlorate of potassa. A portion of the latter is decomposed by the sulphuric acid into oxide of chlorine, bisulphate of potassa, and perchlorate of potassa. The oxide of chlorine inflames the sulphuret of antimony, which is a combustible body, and the whole mass instantly bursts into flame.

same zinc, and make an experiment a little more closely at home, as it were. You will have here the same thing happening. Here is the piece of zinc; there [pointing to a jet of hydrogen] is the furnace, and we will set to work and try and burn the metal. It glows, you see; there is the combustion; and there is the white substance into which it burns. And so, if I take that flame of hydrogen as the representative of a candle, and show you a substance like zinc burning in the flame, you will see that it was merely during the action of combustion that this substance glowed—while it was kept hot; and if I take a flame of hydrogen and put this white substance from the zinc into it, look how beautifully it glows, and just because it is a solid substance.

I will now take such a flame as I had a moment since, and set free from it the particles of carbon. Here is some camphene, which will burn with a smoke; but if I send these particles of smoke through this pipe into the hydrogen flame you will see they will burn and become luminous, because we heat them a second time. There they are. Those are the particles of carbon reignited a second time. They are those particles which you can easily see by holding a piece of paper behind them, and which, while they are in the flame, are ignited by the heat produced, and, when so ignited, produce this brightness. When the particles are not separated you get no brightness. The flame of coal gas owes its brightness to the separation, during combustion, of these particles of carbon, which are equally in that as in a candle. I can very quickly alter that arrangement. Here, for instance, is a bright flame of gas. Supposing I add so much air to the flame as to cause it all to burn before those particles are set free, I shall not have this brightness; and I can do that in this way: If I place over the jet this wire-gauze cap, as you see, and then light the gas over it, it burns with a non-luminous flame, owing to its having plenty of air mixed with it before it burns; and if I raise the gauze, you see it does not burn below(^{10}). There is plenty of carbon

[10] The "air-burner," which is of such value in the laboratory, owes its advantage to this principle. It consists of a cylindrical metal chimney, covered at the top with a piece of rather coarse iron wire gauze. This is supported over an Argand burner in such a manner that the gas may mix in the chimney with an amount of air sufficient to burn the carbon and hydrogen simultaneously, so that there may be no separation of carbon in the flame with consequent deposition of soot. The flame, being unable to pass through the wire gauze, burns in a steady, nearly invisible manner above.

in the gas; but, because the atmosphere can get to it, and mix with it before it burns, you see how pale and blue the flame is. And if I blow upon a bright gas flame, so as to consume all this carbon before it gets heated to the glowing point, it will also burn blue. [The lecturer illustrated his remarks by blowing on the gas light.] The only reason why I have not the same bright light when I thus blow upon the flame is that the carbon meets with sufficient air to burn it before it gets separated in the flame in a free state. The difference is solely due to the solid particles not being separated before the gas is burnt.

You observe that there are certain products as the result of the combustion of a candle, and that of these products one portion may be considered as charcoal, or soot; that charcoal, when afterward burnt, produces some other product; and it concerns us very much now to ascertain what that other product is. We showed that something was going away; and I want you now to understand how much is going up into the air; and for that purpose we will have combustion on a little larger scale. From that candle ascends heated air, and two or three experiments will show you the ascending current; but, in order to give you a notion of the quantity of matter which ascends in this way, I will make an experiment by which I shall try to imprison some of the products of this combustion. For this purpose I have here what boys call a fire-balloon; I use this fire-balloon merely as a sort of measure of the result of the combustion we are considering; and I am about to make a flame in such an easy and simple manner as shall best serve my present purpose. This plate shall be the "cup," we will so say, of the candle; this spirit shall be our fuel; and I am about to place this chimney over it, because it is better for me to do so than to let things proceed at random. Mr. Anderson will now light the fuel, and here at the top we shall get the results of the combustion. What we get at the top of that tube is exactly the same, generally speaking, as you get from the combustion of a candle; but we do not get a luminous flame here, because we use a substance which is feeble in carbon. I am about to put this balloon—not into action, because that is not my object—but to show you the effect which results from the action of those products which arise from the candle, as they arise here from the furnace. [The balloon was held over the chimney (Fig. 64), when it immediately commenced to fill.] You see how it is disposed to ascend; but we must not let it up; because it might come in contact with those upper gaslights, and that would be very inconvenient. [The upper gas-

lights were turned out at the request of the lecturer, and the balloon was allowed to ascend.] Does not that show you what a large bulk of matter is being evolved? Now there is going through this tube [placing a large glass tube over a candle] all the products of that candle, and you will presently see that the tube will become quite opaque. Suppose I take another candle, and place it under a jar, and then put a light on the other side, just to show you what is going on. You see that the sides of the jar become cloudy, and the light begins to burn feebly. It is the products, you see, which make the light so dim, and this is the same thing which makes the sides of the jar so opaque. If you go home, and take a spoon that has been in the cold air, and hold it over a candle—not so as to soot it—you will find that it becomes dim just as that jar is dim. If you can get a silver dish, or something of that kind, you will make the experiment still better; and now, just to carry your thoughts forward to the time we shall next meet, let me tell you that it is *water* which causes the dimness, and when we next meet I will show you that we can make it, without difficulty, assume the form of a liquid.

Fig. 64

LECTURE III

PRODUCTS: WATER FROM THE COMBUSTION— NATURE OF WATER—A COMPOUND— HYDROGEN

I DARE say you well remember that when we parted we had just mentioned the word "products" from the candle; for when a candle burns we found we were able, by nice adjustment, to get various products from it. There was one substance which was not obtained when the candle was burning properly, which was charcoal or smoke, and there was some other substance that went upward from the flame which did not appear as smoke, but took some other form, and made part of that general current which, ascending from the candle upward, becomes invisible, and escapes. There were also other products to mention. You remember that in that rising current having its origin at the candle we found that one part was condensable against a cold spoon, or against a clean plate, or any other cold thing, and another part was incondensable.

We will first take the condensable part, and examine it, and, strange to say, we find that that part of the product is just water— nothing but water. On the last occasion I spoke of it incidentally, merely saying that water was produced among the condensable products of the candle; but to-day I wish to draw your attention to water, that we may examine it carefully, especially in relation to this subject, and also with respect to its general existence on the surface of the globe.

Now, having previously arranged an experiment for the purpose of condensing water from the products of the candle, my next point will be to show you this water; and perhaps one of the best means that I can adopt for showing its presence to so many at once is to exhibit a very visible action of water, and then to apply that test to what is collected as a drop at the bottom of that vessel. I have here a chemical substance, discovered by Sir Humphry Davy, which has a very energetic action upon water, which I shall use as a test of the presence of water. If I take a little piece of it—it is called potassium, as coming from potash—if I take a little piece of it, and throw it into that basin, you see how it shows the presence of water by lighting up and floating about, burning with a violet flame. I am now going to take away the candle which has been burning beneath the vessel containing ice and salt, and you see a drop of

water—a condensed product of the candle—hanging from the under surface of the dish (Fɪɢ. 65). I will show you that potassium has the same action upon it as upon the water in that basin in the experiment we have just tried. See! it takes fire, and burns in just the same manner. I will take another drop upon this glass slab, and when I put the potassium on to it, you see at once, from its taking fire, that there is water present. Now that water was produced by the candle. In the same manner, if I put this spirit lamp under that jar, you will soon see the latter become damp from the dew which is deposited upon it—that dew being the result of combustion; and I have no doubt you will shortly see, by the drops of water which fall upon the paper below, that there is a good deal of

Fɪɢ. 65

water produced from the combustion of the lamp. I will let it remain, and you can afterward see how much water has been collected. So, if I take a gas lamp, and put any cooling arrangement over it, I shall get water—water being likewise produced from the combustion of gas. Here, in this bottle, is a quantity of water—perfectly pure, distilled water, produced from the combustion of a gas lamp—in no point different from the water that you distill from the river, or ocean, or spring, but exactly the same thing. Water is one individual thing; it never changes. We can add to it by careful adjustment for a little while, or we can take it apart and get other things from it; but water, as water, remains always the same, either in a solid, liquid, or fluid state. Here again [holding another bottle] is some water produced by the combustion of an oil lamp. A pint of oil, when burnt fairly and properly, produces rather more than a pint of water. Here, again, is some water, produced by a rather long experiment, from a wax candle. And so we can go on with almost all combustible substances, and find that if they burn with a flame, as a candle, they produce water. You may make these experiments yourselves: the head of a poker is a very good thing to try with, and if it remains cold long enough over the candle, you may get water condensed in drops on it; or a spoon, or ladle, or any thing else may be used, provided it be clean, and can carry off the heat, and so condense the water.

And now—to go into the history of this wonderful production of water from combustibles, and by combustion—I must first of all tell you that this water may exist in different conditions; and although you may now be acquainted with all its forms, they still require us to give a little attention to them for the present; so that we may perceive how the water, while it goes through its Protean changes, is entirely and absolutely the same thing, whether it is produced from a candle, by combustion, or from the rivers or ocean.

First of all, water, when at the coldest, is ice. Now we philosophers —I hope that I may class you and myself together in this case— speak of water as water, whether it be in its solid, or liquid, or gaseous state—we speak of it chemically as water. Water is a thing compounded of two substances, one of which we have derived from the candle, and the other we shall find elsewhere. Water may occur as ice; and you have had most excellent opportunities lately of seeing this. Ice changes back into water—for we had on our last Sabbath a strong instance of this change by the sad catastrophe which occurred in our own house, as well as in the houses of many of our friends—ice changes back into water when the temperature is raised; water also changes into steam when it is warmed enough. The water which we have here before us is in its densest state[11]; and, although it changes in weight, in condition, in form, and in many other qualities, it still is water; and whether we alter it into ice by cooling, or whether we change it into steam by heat, it increases in volume—in the one case very strangely and powerfully, and in the other case very largely and wonderfully. For instance, I will now take this tin cylinder, and pour a little water into it, and, seeing how much water I pour in, you may easily estimate for yourselves how

[11] Water is in its densest state at a temperature of 39.1° Fahrenheit.

282

high it will rise in the vessel: it will cover the bottom about two inches. I am now about to convert the water into steam for the purpose of showing to you the different volumes which water occupies in its different states of water and steam.

Let us now take the case of water changing into ice: we can effect that by cooling it in a mixture of salt and pounded ice([12])—and I shall do so to show you the expansion of water into a thing of larger bulk when it is so changed. These bottles [holding one] are made of strong cast iron, very strong and very thick—I suppose they are the third of an inch in thickness; they are very carefully filled with water, so as to exclude all air, and then they are screwed down tight. We shall see that when we freeze the water in these iron vessels, they will not be able to hold the ice, and the expansion within them will break them in pieces as these [pointing to some fragments] are broken, which have been bottles of exactly the same kind. I am about to put these two bottles into that mixture of ice and salt for the purpose of showing that when water becomes ice it changes in volume in this extraordinary way.

In the mean time, look at the change which has taken place in the water to which we have applied heat; it is losing its fluid state. You may tell this by two or three circumstances. I have covered the mouth of this glass flask, in which water is boiling, with a watch-glass. Do you see what happens? It rattles away like a valve chattering, because the steam rising from the boiling water sends the valve up and down, and forces itself out, and so makes it clatter. You can very easily perceive that the flask is quite full of steam, or else it would not force its way out. You see, also, that the flask contains a substance very much larger than the water, for it fills the whole of the flask over and over again, and there it is blowing away into the air; and yet you can not observe any great diminution in the bulk of the water, which shows you that its change of bulk is very great when it becomes steam.

I have put our iron bottles containing water into this freezing mixture, that you may see what happens. No communication will take place, you observe, between the water in the bottles and the ice in

[12] A mixture of salt and pounded ice reduces the temperature from 32° F. to zero, the ice at the same time becoming fluid.

the outer vessel. But there will be a conveyance of heat from the one to the other, and if we are successful—we are making our experiment in very great haste—I expect you will by-and-by, so soon as the cold has taken possession of the bottles and their contents, hear a pop on the occasion of the bursting of the one bottle or the other, and, when we come to examine the bottles, we shall find their contents masses of ice, partly inclosed by the covering of iron which is too small for them, because the ice is larger in bulk than the

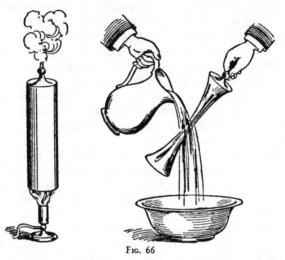

FIG. 66

water. You know very well that ice floats upon water; if a boy falls through a hole into the water, he tries to get on the ice again to float him up. Why does the ice float? Think of that, and philoso-phize. Because the ice is larger than the quantity of water which can produce it, and therefore the ice weighs the lighter and the water is the heavier.

To return now to the action of heat on water. See what a stream of vapor is issuing from this tin vessel! You observe, we must have made it quite full of steam to have it sent out in that great quantity. And now, as we can convert the water into steam by heat, we con-vert it back into liquid water by the application of cold. And if we take a glass, or any other cold thing, and hold it over this steam, see how soon it gets damp with water: it will condense it until the glass is warm—it condenses the water which is now running down the sides of it. I have here another experiment to show the condensation of water from a vaporous state back into a liquid state, in the same way as the vapor, one of the products of the candle, was condensed against the bottom of the dish and obtained in the form of water; and to show you how truly and thoroughly these changes take place, I will take this tin flask, which is now full of steam, and close the top. We shall see what takes place when we cause this water or steam to return back to the fluid state by pouring some cold water on the outside. [The lecturer poured the cold water over the vessel, when it immediately collapsed (Fig. 66).] You see what has happened. If I had closed the stopper, and still kept the heat applied to it, it would have burst the vessel; yet, when the steam returns to the state of water, the vessel collapses, there being a vacuum produced inside by the condensation of the steam. I show you these experiments for the purpose of pointing out that in all these occurrences there is nothing that changes the water into any other thing; it still remains water; and so the vessel is obliged to give way, and is crushed inward, as in the other case, by the farther application of heat, it would have been blown outward.

And what do you think the bulk of that water is when it assumes the vaporous condition? You see that cube [pointing to a cubic foot]. There, by its side, is a cubic inch (Fig. 67), exactly the same shape as the cubic foot, and that bulk of water [the cubic inch] is sufficient to expand into that bulk [the cubic foot] of steam; and, on the contrary, the application of cold will contract that large quantity of steam into this small quantity of water. [One of the iron bottles burst at that moment.] Ah! There is one of our bottles burst, and here, you see, is a crack down one side an eighth of an inch in width. [The other now exploded, sending the freezing mixture in all directions.] This other bottle is also broken; although

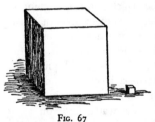

Fig. 67

the iron was nearly half an inch thick, the ice has burst it asunder. These changes always take place in water; they do not require to be always produced by artificial means; we only use them here because we want to produce a small winter round that little bottle instead of a long and severe one. But if you go to Canada, or to the North, you will find the temperature there out of doors will do the same thing as has been done here by the freezing mixture.

To return to our quiet philosophy. We shall not in future be deceived, therefore, by any changes that are produced in water. Water is the same every where, whether produced from the ocean or from the flame of the candle. Where, then, is this water which we get from a candle? I must anticipate a little, and tell you. It evidently comes, as to part of it, from the candle, but is it within the candle beforehand? No, it is not in the candle; and it is not in the air around about the candle which is necessary for its combustion. It is neither in one nor the other, but it comes from their conjoint action, a part from the candle, a part from the air; and this we have now to trace, so that we may understand thoroughly what is the chemical history of a candle when we have it burning on our table. How shall we get at this? I myself know plenty of ways, but I want *you* to get at it from the association in your own minds of what I have already told you.

I think you can see a little in this way. We had just now the case of a substance which acted upon the water in the way that Sir Humphry Davy showed us([13]), and which I am now going to recall to your minds by making again an experiment upon that dish. It is a thing which we have to handle very carefully; for you see, if I allow a little splash of water to come upon this mass, it sets fire to part of it; and if there were free access of air, it would quickly set fire to the whole. Now this is a metal—a beautiful and bright metal—which rapidly changes in the air, and, as you know, rapidly changes in water. I will put a piece on the water, and you see it burns beautifully, making a floating lamp, using the water in the place of air. Again, if we take a few iron filings or turnings and

[13] Potassium, the metallic basis of potash, was discovered by Sir Humphry Davy in 1807, who succeeded in separating it from potash by means of a powerful voltaic battery. Its great affinity for oxygen causes it to decompose water with evolution of hydrogen, which takes fire with the heat produced.

put them in water, we find that they likewise undergo an alteration. They do not change so much as this potassium does, but they change somewhat in the same way; they become rusty, and show an action upon the water, though in a different degree of intensity to what this beautiful metal does; but they act upon the water in the same manner generally as this potassium. I want you to put these different facts together in your minds. I have another metal here [zinc], and when we examined it with regard to the solid substance produced by its combustion, we had an opportunity of seeing that it burned; and I suppose, if I take a little strip of this zinc and put it over the candle, you will see something half way, as it were, between the combustion of potassium on the water and the action of iron— you see there is a sort of combustion. It has burned, leaving a white ash or residuum; and here also we find that the metal has a certain amount of action upon water.

By degrees we have learned how to modify the action of these different substances, and to make them tell us what we want to know. And now, first of all, I take iron. It is a common thing in all chemical reactions, where we get any result of this kind, to find that it is increased by the action of heat; and if we want to examine minutely and carefully the action of bodies one upon another, we often have to refer to the action of heat. You are aware, I believe, that iron filings burn beautifully in the air; but I am about to show you an experiment of this kind, because it will impress upon you what I am going to say about iron in its action on water. If I take a flame and make it hollow—you know why, because I want to get air to it and into it, and therefore I make it hollow—and then take a few iron filings and drop them into the flame, you see how well they burn. That combustion results from the chemical action which is going on when we ignite those particles. And so we proceed to consider these different effects, and ascertain what iron will do when it meets with water. It will tell us the story so beautifully, so gradually and regularly, that I think it will please you very much.

I have here a furnace with a pipe going through it like an iron gun barrel (FIG. 68), and I have stuffed that barrel full of bright iron turnings, and placed it across the fire to be made red-hot. We can either send air through the barrel to come in contact with the iron, or we can send steam from this little boiler at the end of the barrel. Here is a stop-cock which shuts off the steam from the barrel until we wish to admit it. There is some water in these glass jars, which I have colored blue, so that you may see what happens. Now you know very well that any steam I might send through that barrel, if it went through into the water, would be condensed; for you have seen that steam can not retain its gaseous form if it be cooled down; you saw it here [pointing to the tin flask] crushing itself into a small bulk, and causing the flask holding it to collapse;

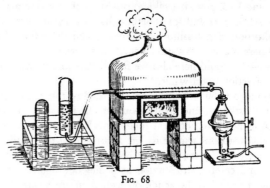

Fig. 68

so that if I were to send steam through that barrel it would be condensed, supposing the barrel were cold; it is, therefore, heated to perform the experiment I am now about to show you. I am going to send the steam through the barrel in small quantities, and you shall judge for yourselves, when you see it issue from the other end, whether it still remains steam. Steam is condensible into water, and when you lower the temperature of steam you convert it back into fluid water; but I have lowered the temperature of the gas which I have collected in this jar by passing it through water after it has traversed the iron barrel, and still it does not change back into water. I will take another test and apply to this gas. (I hold the jar in an inverted position, or my substance would escape.) If I now apply a light to the mouth of the jar, it ignites with a slight noise. That tells you that it is not steam; steam puts out a fire: it does not burn; but you saw that what I had in that jar burnt. We may obtain this substance equally from water produced from the candle flame as from any other source. When it is obtained by the action of the iron upon the aqueous vapor, it leaves the iron in a state very similar to that in which these filings were after they were burnt. It makes the iron heavier than it was before. So long as the iron remains in the tube and is heated, and is cooled again without the access of air or water, it does not change in its weight; but after having had this current of steam passed over it, it then comes out heavier than it was before, having taken something out of the steam, and having allowed something else to pass forth, which we see here. And now, as we have another jar full, I will show you something most interesting. It is a combustible gas; and I might at once take this jar and set fire to the contents, and show you that it is combustible; but I intend to show you more, if I can. It is also a very light substance. Steam will condense; this body will rise in the air, and not condense. Suppose I take another glass jar, empty of all but air: if I examine it with a taper I shall find that it contains nothing but air. I will now take this jar full of the gas that I am speaking of, and deal with it as though it were a light body;

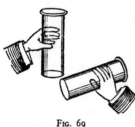

I will hold both upside down, and turn the one up under the other (FIG. 69); and that which did contain the gas procured from the steam, what does it contain now? You will find it now only contains air. But look! Here is the combustible substance [taking the other jar] which I have poured out of the one jar into the other. It still preserves its qual-

FIG. 69

ity, and condition, and independence, and therefore is the more worthy of our consideration, as belonging to the products of a candle.

Now this substance which we have just prepared by the action of iron on the steam or water, we can also get by means of those other things which you have already seen act so well upon the water. If I take a piece of potassium, and make the necessary arrangements, it will produce this gas; and if, instead, a piece of zinc, I find, when I come to examine it very carefully, that the main reason why this zinc can not act upon the water continuously as the other metal does, is because the result of the action of the water envelops the zinc in a kind of protecting coat. We have learned in consequence, that if we put into our vessel only the zinc and water, they, by themselves, do not give rise to much action, and we get no result. But suppose I proceed to dissolve off this varnish—this encumbering substance—which I can do by a little acid; the moment I do this I find the zinc acting upon the water exactly as the iron did, but at the common temperature. The acid in no way is altered, except in its combination with the oxide of zinc which is produced. I have now poured the acid into the glass, and the effect is as though I were applying heat to cause this boiling up. There is something coming off from the zinc very abundantly, which is not steam. There is a jar full of it; and you will find that I have exactly the same combustible substance remaining in the vessel, when I hold it upside down, that I produced during the experiment with the iron barrel. This is what we get from water, the same substance which is contained in the candle.

Let us now trace distinctly the connection between these two points. This is hydrogen—a body classed among those things which

Fig. 70

in chemistry we call elements, because we can get nothing else out of them. A candle is not an elementary body, because we can get carbon out of it; we can get this hydrogen out of it, or at least out of the water which it supplies. And this gas has been so named hydrogen, because it is that element which, in association with another, generates water.* Mr. Anderson having now been able to get two or three jars of gas, we shall have a few experiments to make, and I want to show you the best way of making these experi-

* Ύδωρ, "water," and γεννάω, "I generate."

290

ments. I am not afraid to show you, for I wish you to make experiments, if you will only make them with care and attention, and the assent of those around you. As we advance in chemistry we are obliged to deal with substances which are rather injurious if in their wrong places; the acids, and heat, and combustible things we use, might do harm if carelessly employed. If you want to make hydrogen, you can make it easily from bits of zinc, and sulphuric or muriatic acid. Here is what in former times was called the "philosopher's candle." It is a little phial with a cork and a tube or pipe passing through it. And I am now putting a few little pieces of zinc into it. This little instrument I am going to apply to a useful purpose in our demonstrations, for I want to show you that you can prepare hydrogen, and make some experiments with it as you please, at your own homes. Let me here tell you why I am so careful to fill this phial nearly, and yet not quite full. I do it because the evolved gas, which, as you have seen, is very combustible, is explosive to a considerable extent when mixed with air, and might lead to harm if you were to apply a light to the end of that pipe before all the air had been swept out of the space above the water. I am now about to pour in the sulphuric acid. I have used very little zinc and more sulphuric acid and water, because I want to keep it at work for some time. I therefore take care in this way to modify the proportions of the ingredients so that I may have a regular supply—not too quick and not too slow. Supposing I now take a glass and put it upside down over the end of the tube, because the hydrogen is light I expect that it will remain in that vessel a little while. We will now test the contents of our glass to see if there be hydrogen in it; I think I am safe in saying we have caught some [applying a light]. There it is, you see. I will now apply a light to the top of the tube. There is the hydrogen burning (FIG. 71). There is our philosophical candle. It is a foolish, feeble sort of a flame, you may say, but it is so hot that scarcely any common flame gives out so much heat. It goes on burning regularly, and I am now about to put that flame to burn under a certain arrangement, in order that we may examine its results and make use of the information which we may thereby acquire. Inas-

FIG. 71

much as the candle produces water, and this gas comes out of the water, let us see what this gives us by the same process of combustion that the candle went through when it burnt in the atmosphere; and for that purpose I am going to put the lamp under this apparatus (FIG. 72), in order to condense whatever may arise from the combustion within it. In the course of a short time you will see moisture appearing in the cylinder, and you will get the water running down the side, and the water from this hydrogen flame will have absolutely the same effect upon all our tests, being obtained by the same general process as in the former case. This hydrogen is a very beautiful substance. It is so light that it carries things up; it is far lighter than the atmosphere; FIG. 72 and I dare say I can show you this by an experiment which, if you are very clever, some of you may even have skill enough to repeat. Here is our generator of hydrogen, and here are some soapsuds. I have an India-rubber tube connected with the hydrogen generator, and at the end of the tube is a tobacco pipe. I can thus put the pipe into the suds and blow bubbles by means of the hydrogen. You observe how the bubbles fall downward when I blow them with my warm breath; but notice the difference when I blow them with hydrogen. [The lecturer here blew bubbles with hydrogen, which rose to the roof of the theatre.] It shows you how light this gas must be in order to carry with it not merely the ordinary soap bubble, but the larger portion of a drop hanging to the bottom of it. I can show its lightness in a better way than this; larger bubbles than these may be so lifted up; indeed, in former times balloons used to be filled with this gas. Mr. Anderson will fasten this tube on to our generator, and we shall have a stream of hydrogen here with which we can charge this balloon made of collodion. I need not even be very careful to get all the air out, for I know the power of this gas to carry it up. [Two collodion balloons were inflated and sent up, one being held by a string.] Here is another larger one, made of thin membrane, which we will fill and allow to ascend; you will see they will all remain floating about until the gas escapes.

292

What, then, are the comparative weights of these substances? I have a table here which will show you the proportion which their weights bear to each other. I have taken a pint and a cubic foot as the measures, and have placed opposite to them the respective figures. A pint measure of this hydrogen weighs three-quarters of our smallest weight, a grain, and a cubic foot weighs one-twelfth of an ounce; whereas a pint of water weighs 8,750 grains, and a cubic foot of water weighs almost 1,000 ounces. You see, therefore, what a vast difference there is between the weight of a cubic foot of water and a cubic foot of hydrogen.

Hydrogen gives rise to no substance that can become solid, either during combustion or afterward as a product of its combustion; but when it burns it produces water only; and if we take a cold glass and put it over the flame, it becomes damp, and you have water produced immediately in appreciable quantity; and nothing is produced by its combustion but the same water which you have seen the flame of the candle produce. It is important to remember that this hydrogen is the only thing in nature which furnishes water as the sole product of combustion.

And now we must endeavor to find some additional proof of the general character and composition of water, and for this purpose I will keep you a little longer, so that at our next meeting we may be better prepared for the subject. We have the power of arranging the zinc which you have seen acting upon the water by the assistance of an acid, in such a manner as to cause all the power to be evolved in the place where we require it. I have behind me a voltaic pile, and I am just about to show you, at the end of this lecture, its character and power, that you may see what we shall have to deal with when next we meet. I hold here the extremities of the wires which transport the power from behind me, and which I shall cause to act on the water.

We have previously seen what a power of combustion is possessed by the potassium, or the zinc, or the iron filings; but none of them show such energy as this. [The lecturer here made contact between the two terminal wires of the battery, when a brilliant flash of light was produced.] This light is, in fact, produced by a forty-zinc power of burning; it is a power that I can carry about in my hands through these wires at pleasure, although if I applied it wrongly to myself, it would destroy me in an instant, for it is a most intense thing, and the power you see here put forth while you count five [bringing the poles in contact and exhibiting the electric light] is equivalent to the power of several thunder-storms, so great is its force([14]). And that you may see what intense energy it has, I will take the ends of the wires which convey the power from the battery, and with it I dare say I can burn this iron file. Now this is a chemical power, and one which, when we next meet, I shall apply to water, and show you what results we are able to produce.

[14] Professor Faraday has calculated that there is as much electricity required to decompose one grain of water as there is in a very powerful flash of lightning.

7

Albert Einstein,   THE SPECIAL AND GENERAL THEORY OF
                   RELATIVITY

SCIENCE AND RELIGION

1. Compare Einstein's view of space and time with that
    of Newton. Does the rigor used by Einstein in de-
    scribing exactly what is meant by distance seem
    necessary to you?

2. What does Einstein mean by the principle of rela-
    tivity?

3. Explain, in your own words, the conflict between
    the principle of relativity and the law of propa-
    gation of light. How does Einstein choose to re-
    solve the conflict?

4. The implications of the special theory of relativi-
    ty - that time and distance are not absolute quan-
    tities - seem to be in conflict with our precon-
    ceived notions of time and space. Why should such
    a deep-seated conflict exist?

5. "Everything is relative" is often used as a justi-
    fication for almost any failure to adhere to a set
    of standards and Einstein is cited as the source
    for this quote. Do you believe that this is fair
    use of his work? Explain.

6. How does Einstein resolve the apparent conflict be-
    tween science and religion? Does his argument shed
    any light on the current evolution-creation argu-
    ment?

        Albert Einstein (1879-1955) captured the public
imagination as has no other scientist. "The Gentle
Genius", he not only changed the way we think about
space and time, but also set in motion a second revolution in twentieth-century physics, the quantum theory.

PART I

THE SPECIAL THEORY OF RELATIVITY

I

PHYSICAL MEANING OF GEOMETRICAL PROPOSITIONS

IN your schooldays most of you who read this book made acquaintance with the noble building of Euclid's geometry, and you remember—perhaps with more respect than love—the magnificent structure, on the lofty staircase of which you were chased about for uncounted hours by conscientious teachers. By reason of your past experience, you would certainly regard everyone with disdain who should pronounce even the most out-of-the-way proposition of this science to be untrue. But perhaps this feeling of proud certainty would leave you immediately if some one were to ask you : "What, then, do you mean by the assertion that these propositions are true ? " Let us proceed to give this question a little consideration.

Geometry sets out from certain conceptions such as "plane," "point," and "straight line," with which we are able to associate more or less definite ideas, and from certain simple propositions (axioms) which, in virtue of these ideas, we are inclined to accept as " true." Then, on the basis of a logical process, the justification of which we feel ourselves compelled to admit, all remaining propositions are shown to follow from those axioms, *i.e.* they are proven. A proposition is then correct (" true ") when it has been derived in the recognised manner from the axioms. The question of the " truth " of the individual geometrical propositions is thus reduced to one of the " truth " of the axioms. Now it has long been known that the last question is not only unanswerable by the methods of geometry, but that it is in itself entirely without meaning. We cannot ask whether it is true that only one straight line goes through two points. We can only say that Euclidean geometry deals with things called " straight lines," to each of which is ascribed the property of being uniquely determined by two points situated on it. The concept " true " does not tally with the assertions of pure geometry, because by the word " true " we are eventually in the habit of designating always the correspondence with a " real " object ; geometry, however, is not concerned with the relation of the ideas involved in it to objects of experience, but only with the logical connection of these ideas among themselves.

It is not difficult to understand why, in spite of this, we feel constrained to call the propositions of geometry " true." Geometrical ideas correspond to more or less exact objects in nature, and these last are undoubtedly the exclusive cause of the genesis of those ideas. Geometry ought to refrain from such a course, in order to give to its structure the largest possible logical unity. The practice, for example, of seeing in a "distance" two marked positions on a practically rigid body is something which is lodged deeply in our habit of thought. We are accustomed further to regard three points as being situated on a straight line, if their apparent positions can be made to coincide for observation with one eye, under suitable choice of our place of observation.

If, in pursuance of our habit of thought, we now supplement the propositions of Euclidean geometry by the single proposition that two points on a practically rigid body always correspond to the same distance (line-interval), independently of any changes in position to which we may subject the body, the propositions of Euclidean geometry then resolve themselves into propositions on the possible relative position of practically rigid bodies.[1] Geometry which has been supplemented in this way is then to be treated as a branch of physics. We can now legitimately ask as to the "truth" of geometrical propositions interpreted in this way, since we are justified in asking whether these propositions are satisfied for those real things we have associated with the geometrical ideas. In less exact terms we can express this by saying that by the "truth" of a geometrical proposition in this sense we understand its validity for a construction with ruler and compasses.

---

[1] It follows that a natural object is associated also with a straight line. Three points A, B and C on a rigid body thus lie in a straight line when, the points A and C being given, B is chosen such that the sum of the distances AB and BC is as short as possible. This incomplete suggestion will suffice for our present purpose.

SPECIAL THEORY OF RELATIVITY

Of course the conviction of the " truth " of geometrical propositions in this sense is founded exclusively on rather incomplete experience. For the present we shall assume the " truth " of the geometrical propositions, then at a later stage (in the general theory of relativity) we shall see that this " truth " is limited, and we shall consider the extent of its limitation.

II

THE SYSTEM OF CO-ORDINATES

ON the basis of the physical interpretation of distance which has been indicated, we are also in a position to establish the distance between two points on a rigid body by means of measurements. For this purpose we require a " distance " (rod S) which is to be used once and for all, and which we employ as a standard measure. If, now, A and B are two points on a rigid body, we can construct the line joining them according to the rules of geometry ; then, starting from A, we can mark off the distance S time after time until we reach B. The number of these operations required is the numerical measure of the distance AB. This is the basis of all measurement of length.[1]

Every description of the scene of an event or of the position of an object in space is based on the specification of the point on a rigid body (body of reference) with which that event or object coincides. This applies not only to scientific description, but also to everyday life. If I analyse the place specification " Trafalgar

---

[1] Here we have assumed that there is nothing left over *i.e.* that the measurement gives a whole number. This difficulty is got over by the use of divided measuring-rods, the introduction of which does not demand any fundamentally new method.

Square, London," [1] I arrive at the following result. The earth is the rigid body to which the specification of place refers; "Trafalgar Square, London," is a well-defined point, to which a name has been assigned, and with which the event coincides in space. [2]

This primitive method of place specification deals only with places on the surface of rigid bodies, and is dependent on the existence of points on this surface which are distinguishable from each other. But we can free ourselves from both of these limitations without altering the nature of our specification of position. If, for instance, a cloud is hovering over Trafalgar Square, then we can determine its position relative to the surface of the earth by erecting a pole perpendicularly on the Square, so that it reaches the cloud. The length of the pole measured with the standard measuring-rod, combined with the specification of the position of the foot of the pole, supplies us with a complete place specification. On the basis of this illustration, we are able to see the manner in which a refinement of the conception of position has been developed.

(a) We imagine the rigid body, to which the place specification is referred, supplemented in such a manner that the object whose position we require is reached by the completed rigid body.

(b) In locating the position of the object, we make use of a number (here the length of the pole measured

---

[1] I have chosen this as being more familiar to the English reader than the " Potsdamer Platz, Berlin," which is referred to in the original. (R. W. L.)

[2] It is not necessary here to investigate further the significance of the expression " coincidence in space." This conception is sufficiently obvious to ensure that differences of opinion are scarcely likely to arise as to its applicability in practice.

with the measuring-rod) instead of designated points of reference.

(c) We speak of the height of the cloud even when the pole which reaches the cloud has not been erected. By means of optical observations of the cloud from different positions on the ground, and taking into account the properties of the propagation of light, we determine the length of the pole we should have required in order to reach the cloud.

From this consideration we see that it will be advantageous if, in the description of position, it should be possible by means of numerical measures to make ourselves independent of the existence of marked positions (possessing names) on the rigid body of reference. In the physics of measurement this is attained by the application of the Cartesian system of co-ordinates.

This consists of three plane surfaces perpendicular to each other and rigidly attached to a rigid body. Referred to a system of co-ordinates, the scene of any event will be determined (for the main part) by the specification of the lengths of the three perpendiculars or co-ordinates (x, y, z) which can be dropped from the scene of the event to those three plane surfaces. The lengths of these three perpendiculars can be determined by a series of manipulations with rigid measuring-rods performed according to the rules and methods laid down by Euclidean geometry.

In practice, the rigid surfaces which constitute the system of co-ordinates are generally not available ; furthermore, the magnitudes of the co-ordinates are not actually determined by constructions with rigid rods, but by indirect means. If the results of physics and astronomy are to maintain their clearness, the physical meaning of specifications of position must always be sought in accordance with the above considerations.[1]

We thus obtain the following result : Every description of events in space involves the use of a rigid body to which such events have to be referred. The resulting relationship takes for granted that the laws of Euclidean geometry hold for " distances," the " distance " being represented physically by means of the convention of two marks on a rigid body.

---

[1] A refinement and modification of these views does not become necessary until we come to deal with the general theory of relativity, treated in the second part of this book.

SPACE AND TIME IN CLASSICAL MECHANICS

THE purpose of mechanics is to describe how bodies change their position in space with "time." I should load my conscience with grave sins against the sacred spirit of lucidity were I to formulate the aims of mechanics in this way, without serious reflection and detailed explanations. Let us proceed to disclose these sins.

It is not clear what is to be understood here by "position" and "space." I stand at the window of a railway carriage which is travelling uniformly, and drop a stone on the embankment, without throwing it. Then, disregarding the influence of the air resistance, I see the stone descend in a straight line. A pedestrian who observes the misdeed from the footpath notices that the stone falls to earth in a parabolic curve. I now ask: Do the "positions" traversed by the stone lie "in reality" on a straight line or on a parabola? Moreover, what is meant here by motion "in space"? From the considerations of the previous section the answer is self-evident. In the first place we entirely shun the vague word "space," of which, we must honestly acknowledge, we cannot form the slightest conception, and we replace it by "motion relative to a practically rigid body of reference." The positions relative to the body of reference (railway carriage or embankment) have already been defined in detail in the preceding section. If instead of " body of reference " we insert " system of co-ordinates," which is a useful idea for mathematical description, we are in a position to say : The stone traverses a straight line relative to a system of co-ordinates rigidly attached to the carriage, but relative to a system of co-ordinates rigidly attached to the ground (embankment) it describes a parabola. With the aid of this example it is clearly seen that there is no such thing as an independently existing trajectory (lit. " path-curve " [1]), but only a trajectory relative to a particular body of reference.

In order to have a *complete* description of the motion, we must specify how the body alters its position *with time* ; *i.e.* for every point on the trajectory it must be stated at what time the body is situated there. These data must be supplemented by such a definition of time that, in virtue of this definition, these time-values can be regarded essentially as magnitudes (results of measurements) capable of observation. If we take our stand on the ground of classical mechanics, we can satisfy this requirement for our illustration in the following manner. We imagine two clocks of identical construction ; the man at the railway-carriage window is holding one of them, and the man on the foot-path the other. Each of the observers determines the position on his own reference-body occupied by the stone at each tick of the clock he is holding in his hand. In this connection we have not taken account of the inaccuracy involved by the finiteness of the velocity of propagation of light. With this and with a second difficulty prevailing here we shall have to deal in detail later.

[1] That is, a curve along which the body moves.

IV

THE GALILEIAN SYSTEM OF CO-ORDINATES

AS is well known, the fundamental law of the mechanics of Galilei-Newton, which is known as the *law of inertia*, can be stated thus : A body removed sufficiently far from other bodies continues in a state of rest or of uniform motion in a straight line. This law not only says something about the motion of the bodies, but it also indicates the reference-bodies or systems of co-ordinates, permissible in mechanics, which can be used in mechanical description. The visible fixed stars are bodies for which the law of inertia certainly holds to a high degree of approximation. Now if we use a system of co-ordinates which is rigidly attached to the earth, then, relative to this system, every fixed star describes a circle of immense radius in the course of an astronomical day, a result which is opposed to the statement of the law of inertia. So that if we adhere to this law we must refer these motions only to systems of co-ordinates relative to which the fixed stars do not move in a circle. A system of co-ordinates of which the state of motion is such that the law of inertia holds relative to it is called a " Galileian system of co-ordinates." The laws of the mechanics of Galilei-Newton can be regarded as valid only for a Galileian system of co-ordinates.

V

THE PRINCIPLE OF RELATIVITY (IN THE RESTRICTED SENSE)

IN order to attain the greatest possible clearness, let us return to our example of the railway carriage supposed to be travelling uniformly. We call its motion a uniform translation ("uniform" because it is of constant velocity and direction, "translation" because although the carriage changes its position relative to the embankment yet it does not rotate in so doing). Let us imagine a raven flying through the air in such a manner that its motion, as observed from the embankment, is uniform and in a straight line. If we were to observe the flying raven from the moving railway carriage, we should find that the motion of the raven would be one of different velocity and direction, but that it would still be uniform and in a straight line. Expressed in an abstract manner we may say: If a mass m is moving uniformly in a straight line with respect to a co-ordinate system K, then it will also be moving uniformly and in a straight line relative to a second co-ordinate system K', provided that the latter is executing a uniform translatory motion with respect to K. In accordance with the discussion contained in the preceding section, it follows that:

THE PRINCIPLE OF RELATIVITY

If K is a Galileian co-ordinate system, then every other co-ordinate system K' is a Galileian one, when, in relation to K, it is in a condition of uniform motion of translation. Relative to K' the mechanical laws of Galilei-Newton hold good exactly as they do with respect to K.

We advance a step farther in our generalisation when we express the tenet thus : If, relative to K, K' is a uniformly moving co-ordinate system devoid of rotation, then natural phenomena run their course with respect to K' according to exactly the same general laws as with respect to K. This statement is called the *principle of relativity* (in the restricted sense).

As long as one was convinced that all natural phenomena were capable of representation with the help of classical mechanics, there was no need to doubt the validity of this principle of relativity. But in view of the more recent development of electrodynamics and optics it became more and more evident that classical mechanics affords an insufficient foundation for the physical description of all natural phenomena. At this juncture the question of the validity of the principle of relativity became ripe for discussion, and it did not appear impossible that the answer to this question might be in the negative.

Nevertheless, there are two general facts which at the outset speak very much in favour of the validity of the principle of relativity. Even though classical mechanics does not supply us with a sufficiently broad basis for the theoretical presentation of all physical phenomena, still we must grant it a considerable measure of " truth," since it supplies us with the actual motions of the heavenly bodies with a delicacy of detail little short of wonderful. The principle of relativity must therefore apply with great accuracy in the domain of *mechanics*. But that a principle of such broad generality should hold with such exactness in one domain of phenomena, and yet should be invalid for another, is *a priori* not very probable.

We now proceed to the second argument, to which, moreover, we shall return later. If the principle of relativity (in the restricted sense) does not hold, then the Galileian co-ordinate systems K, K', K'', etc., which are moving uniformly relative to each other, will not be *equivalent* for the description of natural phenomena. In this case we should be constrained to believe that natural laws are capable of being formulated in a particularly simple manner, and of course only on condition that, from amongst all possible Galileian co-ordinate systems, we should have chosen *one* (K_0) of a particular state of motion as our body of reference. We should then be justified (because of its merits for the description of natural phenomena) in calling this system " absolutely at rest," and all other Galileian systems K " in motion." If, for instance, our embankment were the system K_0, then our railway carriage would be a system K, relative to which less simple laws would hold than with respect to K_0. This diminished simplicity would be due to the fact that the carriage K would be in motion (*i.e.* " really ") with respect to K_0. In the general laws of nature which have been formulated with reference to K, the magnitude and direction of the velocity of the carriage would necessarily play a part. We should expect, for instance, that the note emitted by an organ-pipe placed with its axis parallel to the direction of travel would be different from that emitted if the axis of the pipe were placed perpendicular to this direction.

THE PRINCIPLE OF RELATIVITY

Now in virtue of its motion in an orbit round the sun, our earth is comparable with a railway carriage travelling with a velocity of about 30 kilometres per second. If the principle of relativity were not valid we should therefore expect that the direction of motion of the earth at any moment would enter into the laws of nature, and also that physical systems in their behaviour would be dependent on the orientation in space with respect to the earth. For owing to the alteration in direction of the velocity of revolution of the earth in the course of a year, the earth cannot be at rest relative to the hypothetical system K_0 throughout the whole year. However, the most careful observations have never revealed such anisotropic properties in terrestrial physical space, *i.e.* a physical non-equivalence of different directions. This is very powerful argument in favour of the principle of relativity.

VI

THE THEOREM OF THE ADDITION OF VELOCITIES EMPLOYED IN CLASSICAL MECHANICS

LET us suppose our old friend the railway carriage to be travelling along the rails with a constant velocity v, and that a man traverses the length of the carriage in the direction of travel with a velocity w. How quickly or, in other words, with what velocity W does the man advance relative to the embankment during the process ? The only possible answer seems to result from the following consideration : If the man were to stand still for a second, he would advance relative to the embankment through a distance v equal numerically to the velocity of the carriage. As a consequence of his walking, however, he traverses an additional distance w relative to the carriage, and hence also relative to the embankment, in this second, the distance w being numerically equal to the velocity with which he is walking. Thus in total he covers the distance $W=v+w$ relative to the embankment in the second considered. We shall see later that this result, which expresses the theorem of the addition of velocities employed in classical mechanics, cannot be maintained ; in other words, the law that we have just written down does not hold in reality. For the time being, however, we shall assume its correctness.

VII

THE APPARENT INCOMPATIBILITY OF THE LAW OF PROPAGATION OF LIGHT WITH THE PRINCIPLE OF RELATIVITY

THERE is hardly a simpler law in physics than that according to which light is propagated in empty space. Every child at school knows, or believes he knows, that this propagation takes place in straight lines with a velocity $c = 300,000$ km./sec. At all events we know with great exactness that this velocity is the same for all colours, because if this were not the case, the minimum of emission would not be observed simultaneously for different colours during the eclipse of a fixed star by its dark neighbour. By means of similar considerations based on observations of double stars, the Dutch astronomer De Sitter was also able to show that the velocity of propagation of light cannot depend on the velocity of motion of the body emitting the light. The assumption that this velocity of propagation is dependent on the direction " in space " is in itself improbable.

In short, let us assume that the simple law of the constancy of the velocity of light c (in vacuum) is justifiably believed by the child at school. Who would imagine that this simple law has plunged the conscientiously thoughtful physicist into the greatest intellectual difficulties ? Let us consider how these difficulties arise.

Of course we must refer the process of the propagation of light (and indeed every other process) to a rigid reference-body (co-ordinate system). As such a system let us again choose our embankment. We shall imagine the air above it to have been removed. If a ray of light be sent along the embankment, we see from the above that the tip of the ray will be transmitted with the velocity c relative to the embankment. Now let us suppose that our railway carriage is again travelling along the railway lines with the velocity v, and that its direction is the same as that of the ray of light, but its velocity of course much less. Let us inquire about the velocity of propagation of the ray of light relative to the carriage. It is obvious that we can here apply the consideration of the previous section, since the ray of light plays the part of the man walking along relatively to the carriage. The velocity W of the man relative to the embankment is here replaced by the velocity of light relative to the embankment. w is the required velocity of light with respect to the carriage, and we have

$$w = c - v.$$

The velocity of propagation of a ray of light relative to the carriage thus comes out smaller than c.

But this result comes into conflict with the principle of relativity set forth in Section V. For, like every other general law of nature, the law of the transmission of light *in vacuo* must, according to the principle of relativity, be the same for the railway carriage as reference-body as when the rails are the body of reference. But, from our above consideration, this would appear to be impossible. If every ray of light is propagated relative to the embankment with the velocity *c*, then for this reason it would appear that another law of propagation of light must necessarily hold with respect to the carriage—a result contradictory to the principle of relativity.

In view of this dilemma there appears to be nothing else for it than to abandon either the principle of relativity or the simple law of the propagation of light *in vacuo*. Those of you who have carefully followed the preceding discussion are almost sure to expect that we should retain the principle of relativity, which appeals so convincingly to the intellect because it is so natural and simple. The law of the propagation of light *in vacuo* would then have to be replaced by a more complicated law conformable to the principle of relativity. The development of theoretical physics shows, however, that we cannot pursue this course. The epoch-making theoretical investigations of H. A. Lorentz on the electrodynamical and optical phenomena connected with moving bodies show that experience in this domain leads conclusively to a theory of electromagnetic phenomena, of which the law of the constancy of the velocity of light *in vacuo* is a necessary consequence. Prominent theoretical physicists were therefore more inclined to reject the principle of relativity, in spite of the fact that no empirical data had been found which were contradictory to this principle.

At this juncture the theory of relativity entered the arena. As a result of an analysis of the physical conceptions of time and space, it became evident that *in reality there is not the least incompatibility between the*

principle of relativity and the law of propagation of light,
and that by systematically holding fast to both these
laws a logically rigid theory could be arrived at. This
theory has been called the *special theory of relativity*
to distinguish it from the extended theory, with which
we shall deal later. In the following pages we shall
present the fundamental ideas of the special theory of
relativity.

VIII

ON THE IDEA OF TIME IN PHYSICS

LIGHTNING has struck the rails on our railway
embankment at two places *A* and *B* far distant
from each other. I make the additional assertion
that these two lightning flashes occurred simultaneously.
If I ask you whether there is sense in this statement,
you will answer my question with a decided
" Yes." But if I now approach you with the request
to explain to me the sense of the statement more
precisely, you find after some consideration that the
answer to this question is not so easy as it appears at
first sight.

After some time perhaps the following answer would
occur to you : " The significance of the statement is
clear in itself and needs no further explanation ; of
course it would require some consideration if I were to
be commissioned to determine by observations whether
in the actual case the two events took place simultaneously or not." I cannot be satisfied with this answer
for the following reason. Supposing that as a result
of ingenious considerations an able meteorologist were
to discover that the lightning must always strike the
places *A* and *B* simultaneously, then we should be faced
with the task of testing whether or not this theoretical
result is in accordance with the reality. We encounter

the same difficulty with all physical statements in which the conception "simultaneous" plays a part. The concept does not exist for the physicist until he has the possibility of discovering whether or not it is fulfilled in an actual case. We thus require a definition of simultaneity such that this definition supplies us with the method by means of which, in the present case, he can decide by experiment whether or not both the lightning strokes occurred simultaneously. As long as this requirement is not satisfied, I allow myself to be deceived as a physicist (and of course the same applies if I am not a physicist), when I imagine that I am able to attach a meaning to the statement of simultaneity. (I would ask the reader not to proceed farther until he is fully convinced on this point.)

After thinking the matter over for some time you then offer the following suggestion with which to test simultaneity. By measuring along the rails, the connecting line AB should be measured up and an observer placed at the mid-point M of the distance AB. This observer should be supplied with an arrangement (*e.g.* two mirrors inclined at 90°) which allows him visually to observe both places A and B at the same time. If the observer perceives the two flashes of lightning at the same time, then they are simultaneous.

I am very pleased with this suggestion, but for all that I cannot regard the matter as quite settled, because I feel constrained to raise the following objection: "Your definition would certainly be right, if only I knew that the light by means of which the observer at M perceives the lightning flashes travels along the length $A \longrightarrow M$ with the same velocity as along the length $B \longrightarrow M$. But an examination of this supposition would only be possible if we already had at our disposal the means of measuring time. It would thus appear as though we were moving here in a logical circle."

After further consideration you cast a somewhat disdainful glance at me—and rightly so—and you declare : " I maintain my previous definition nevertheless, because in reality it assumes absolutely nothing about light. There is only *one* demand to be made of the definition of simultaneity, namely, that in every real case it must supply us with an empirical decision as to whether or not the conception that has to be defined is fulfilled. That my definition satisfies this demand is indisputable. That light requires the same time to traverse the path $A \longrightarrow M$ as for the path $B \longrightarrow M$ is in reality neither a *supposition nor a hypothesis* about the physical nature of light, but a *stipulation* which I can make of my own freewill in order to arrive at a definition of simultaneity."

It is clear that this definition can be used to give an exact meaning not only to *two* events, but to as many events as we care to choose, and independently of the positions of the scenes of the events with respect to the body of reference [1] (here the railway embankment). We are thus led also to a definition of " time " in physics. For this purpose we suppose that clocks of identical construction are placed at the points A, B and C of

---

[1] We suppose further, that, when three events A, B and C occur in different places in such a manner that A is simultaneous with B, and B is simultaneous with C (simultaneous in the sense of the above definition), then the criterion for the simultaneity of the pair of events A, C is also satisfied. This assumption is a physical hypothesis about the law of propagation of light ; it must certainly be fulfilled if we are to maintain the law of the constancy of the velocity of light *in vacuo*.

the railway line (co-ordinate system), and that they are set in such a manner that the positions of their pointers are simultaneously (in the above sense) the same. Under these conditions we understand by the " time " of an event the reading (position of the hands) of that one of these clocks which is in the immediate vicinity (in space) of the event. In this manner a time-value is associated with every event which is essentially capable of observation.

This stipulation contains a further physical hypothesis, the validity of which will hardly be doubted without empirical evidence to the contrary. It has been assumed that all these clocks go *at the same rate* if they are of identical construction. Stated more exactly : When two clocks arranged at rest in different places of a reference-body are set in such a manner that a *particular* position of the pointers of the one clock is *simultaneous* (in the above sense) with the *same* position of the pointers of the other clock, then identical " settings " are always simultaneous (in the sense of the above definition).

IX

THE RELATIVITY OF SIMULTANEITY

UP to now our considerations have been referred to a particular body of reference, which we have styled a "railway embankment." We suppose a very long train travelling along the rails with the constant velocity *v* and in the direction indicated in Fig. 1. People travelling in this train will with advantage use the train as a rigid reference-body (co-ordinate system); they regard all events in

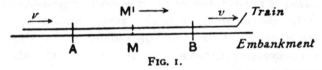

FIG. 1.

reference to the train. Then every event which takes place along the line also takes place at a particular point of the train. Also the definition of simultaneity can be given relative to the train in exactly the same way as with respect to the embankment. As a natural consequence, however, the following question arises:

Are two events (*e.g.* the two strokes of lightning *A* and *B*) which are simultaneous *with reference to the railway embankment* also simultaneous *relatively to the train*? We shall show directly that the answer must be in the negative.

When we say that the lightning strokes *A* and *B* are

318

simultaneous with respect to the embankment, we mean : the rays of light emitted at the places A and B, where the lightning occurs, meet each other at the mid-point M of the length $A \longrightarrow B$ of the embankment. But the events A and B also correspond to positions A and B on the train. Let M' be the mid-point of the distance $A \longrightarrow B$ on the travelling train. Just when the flashes [1] of lightning occur, this point M' naturally coincides with the point M, but it moves towards the right in the diagram with the velocity v of the train. If an observer sitting in the position M' in the train did not possess this velocity, then he would remain permanently at M, and the light rays emitted by the flashes of lightning A and B would reach him simultaneously, *i.e.* they would meet just where he is situated. Now in reality (considered with reference to the railway embankment) he is hastening towards the beam of light coming from B, whilst he is riding on ahead of the beam of light coming from A. Hence the observer will see the beam of light emitted from B earlier than he will see that emitted from A. Observers who take the railway train as their reference-body must therefore come to the conclusion that the lightning flash B took place earlier than the lightning flash A. We thus arrive at the important result :

Events which are simultaneous with reference to the embankment are not simultaneous with respect to the train, and *vice versa* (relativity of simultaneity). Every reference-body (co-ordinate system) has its own particular time ; unless we are told the reference-body to which the statement of time refers, there is no meaning in a statement of the time of an event.

[1] As judged from the embankment.

RELATIVITY OF SIMULTANEITY

Now before the advent of the theory of relativity it had always tacitly been assumed in physics that the statement of time had an absolute significance, *i.e.* that it is independent of the state of motion of the body of reference. But we have just seen that this assumption is incompatible with the most natural definition of simultaneity; if we discard this assumption, then the conflict between the law of the propagation of light *in vacuo* and the principle of relativity (developed in Section VII) disappears.

We were led to that conflict by the considerations of Section VI, which are now no longer tenable. In that section we concluded that the man in the carriage, who traverses the distance *w per second* relative to the carriage, traverses the same distance also with respect to the embankment *in each second* of time. But, according to the foregoing considerations, the time required by a particular occurrence with respect to the carriage must not be considered equal to the duration of the same occurrence as judged from the embankment (as reference-body). Hence it cannot be contended that the man in walking travels the distance *w* relative to the railway line in a time which is equal to one second as judged from the embankment.

Moreover, the considerations of Section VI are based on yet a second assumption, which, in the light of a strict consideration, appears to be arbitrary, although it was always tacitly made even before the introduction of the theory of relativity.

X

ON THE RELATIVITY OF THE CONCEPTION OF DISTANCE

LET us consider two particular points on the train [1] travelling along the embankment with the velocity v, and inquire as to their distance apart. We already know that it is necessary to have a body of reference for the measurement of a distance, with respect to which body the distance can be measured up. It is the simplest plan to use the train itself as reference-body (co-ordinate system). An observer in the train measures the interval by marking off his measuring-rod in a straight line (*e.g.* along the floor of the carriage) as many times as is necessary to take him from the one marked point to the other. Then the number which tells us how often the rod has to be laid down is the required distance.

It is a different matter when the distance has to be judged from the railway line. Here the following method suggests itself. If we call A' and B' the two points on the train whose distance apart is required, then both of these points are moving with the velocity v along the embankment. In the first place we require to determine the points A and B of the embankment which are just being passed by the two points A' and B' at a

[1] *e.g.* the middle of the first and of the twentieth carriage.

particular time t—judged from the embankment. These points A and B of the embankment can be determined by applying the definition of time given in Section VIII. The distance between these points A and B is then measured by repeated application of the measuring-rod along the embankment.

A priori it is by no means certain that this last measurement will supply us with the same result as the first. Thus the length of the train as measured from the embankment may be different from that obtained by measuring in the train itself. This circumstance leads us to a second objection which must be raised against the apparently obvious consideration of Section VI. Namely, if the man in the carriage covers the distance w in a unit of time—*measured from the train,*—then this distance—*as measured from the embankment*—is not necessarily also equal to w.

SCIENCE AND RELIGION

I

D URING THE LAST CENTURY, and part of the one before, it was widely held that there was an unreconcilable conflict between knowledge and belief. The opinion prevailed among advanced minds that it was time that belief should be replaced increasingly by knowledge; belief that did not itself rest on knowledge was superstition, and as such had to be opposed. According to this conception, the sole function of education was to open the way to thinking and knowing, and the school, as the outstanding organ for the people's education, must serve that end exclusively.

One will probably find but rarely, if at all, the rationalistic standpoint expressed in such crass form; for any sensible man would see at once how one-sided is such a statement of the position. But it is just as well to state a thesis starkly and nakedly, if one wants to clear up one's mind as to its nature.

It is true that convictions can best be supported with experience and clear thinking. On this point one must agree unreservedly with the extreme rationalist. The weak point of his conception is, however, this, that those convictions which are necessary and determinant for our conduct and judgments, cannot be found solely along this solid scientific way.

For the scientific method can teach us nothing else beyond how facts are related to, and conditioned by, each other. The aspiration toward such objective knowledge belongs to the highest of which man is capable, and you will certainly not suspect me of wishing to belittle the achievements and the heroic efforts of man in this sphere. Yet it is equally clear that knowledge of what *is* does not open the door directly to what *should be*. One can have the clearest and most complete knowledge of what *is*, and yet not be able to deduct from that what should be the *goal* of our human aspirations. Objective knowledge provides us with powerful instruments for the achievements of certain ends, but the ultimate goal itself and the longing to reach it must come from another source. And it is hardly necessary to argue for the view that our existence and our activity acquire meaning only by the setting up of such a goal and of corresponding values. The knowledge of truth as such is wonderful, but it is so little capable of acting as a guide that it cannot prove even the justification and the value of the aspiration towards that very knowledge of truth. Here we face, therefore, the limits of the purely rational conception of our existence.

But it must not be assumed that intelligent thinking can play no part in the formation of the goal and of ethical judgments. When someone realizes that for the achievement of an end certain means would be useful, the means itself becomes thereby an end. Intelligence makes clear to us the interrelation of means and ends. But mere thinking cannot give us a sense of the ultimate and fundamental ends. To make clear these fundamental ends and valuations, and to set them fast in the emotional life of the individual, seems to me precisely the most important function which religion has to perform in the social life of man. And if one asks whence derives the authority of such fundamental ends, since they cannot be stated and justified merely by reason, one can only answer: they exist in a healthy society as powerful traditions, which act upon the conduct and aspirations and judgments of the individuals; they are there, that is, as something living, without its being necessary to find justification for their existence. They come into being not through demonstration but through revelation, through the medium of powerful personalities. One must not attempt to justify them, but rather to sense their nature simply and clearly.

The highest principles for our aspirations and judgments are given to us in the Jewish-Christian religious tradition. It is a very high goal which, with our weak powers, we can reach only very inadequately, but which gives a sure foundation to our aspirations and valuations. If one were to take that goal out of its religious form and look merely at its purely human side, one might state it perhaps thus: free and responsible development of the individual, so that he may place his powers freely and gladly in the service of all mankind.

There is no room in this for the divinization of a nation, of a class, let alone of an individual. Are we not all children of one father, as it is said in religious language? Indeed, even the divinization of humanity, as an abstract totality, would not be in the spirit of that ideal. It is only to the individual that a soul is given. And the high destiny of the individual is to serve rather than to rule, or to impose himself in any other way.

If one looks at the substance rather than at the form, then one can take these words as expressing also the fundamental democratic position. The true democrat can worship his nation as little as can the man who is religious, in our sense of the term.

What, then, in all this, is the function of education and of the school? They should help the young person to grow up in such a spirit that these fundamental principles should be to him as the air which he breathes. Teaching alone cannot do that.

If one holds these high principles clearly before one's eyes, and compares them with the life and spirit of our times, then it appears glaringly that civilized mankind finds itself at present in grave danger. In the totalitarian states it is the rulers themselves who strive actually to destroy that spirit of hu-

manity. In less threatened parts it is nationalism and intoler-
ance, as well as the oppression of the individuals by economic
means, which threaten to choke these most precious traditions.

A realization of how great is the danger is spreading, however, among thinking people, and there is much search for
means with which to meet the danger—means in the field of
national and international politics, of legislation, of organization in general. Such efforts are, no doubt, greatly needed.
Yet the ancients knew something which we seem to have
forgotten. All means prove but a blunt instrument, if they
have not behind them a living spirit. But if the longing for
the achievement of the goal is powerfully alive within us,
then shall we not lack the strength to find the means for
reaching the goal and for translating it into deeds.

II

It would not be difficult to come to an agreement as to
what we understand by science. Science is the century-old
endeavor to bring together by means of systematic thought
the perceptible phenomena of this world into as thorough-
going an association as possible. To put it boldly, it is the
attempt at the posterior reconstruction of existence by the
process of conceptualization. But when asking myself what
religion is I cannot think of the answer so easily. And even
after finding an answer which may satisfy me at this particular moment I still remain convinced that I can never under
any circumstances bring together, even to a slight extent, all
those who have given this question serious consideration.

At first, then, instead of asking what religion is I should
prefer to ask what characterizes the aspirations of a person
who gives me the impression of being religious: A person
who is religiously enlightened appears to me to be one who
has, to the best of his ability, liberated himself from the

326

fetters of his selfish desires and is preoccupied with thoughts, feelings, and aspirations to which he clings because of their super-personal value. It seems to me that what is important is the force of this super-personal content and the depth of the conviction concerning its overpowering meaningfulness, regardless of whether any attempt is made to unite this content with a divine Being, for otherwise it would not be possible to count Buddha and Spinoza as religious personalities. Accordingly, a religious person is devout in the sense that he has no doubt of the significance and loftiness of those super-personal objects and goals which neither require nor are capable of rational foundation. They exist with the same necessity and matter-of-factness as he himself. In this sense religion is the age-old endeavor of mankind to become clearly and completely conscious of these values and goals and constantly to strengthen and extend their effect. If one conceives of religion and science according to these definitions then a conflict between them appears impossible. For science can only ascertain what *is*, but not what *should be*, and outside of its domain value judgments of all kinds remain necessary. Religion, on the other hand, deals only with evaluations of human thought and action: it cannot justifiably speak of facts and relationships between facts. According to this interpretation the well-known conflicts between religion and science in the past must all be ascribed to a misapprehension of the situation which has been described.

For example, a conflict arises when a religious community insists on the absolute truthfulness of all statements recorded in the Bible. This means an intervention on the part of religion into the sphere of science; this is where the struggle of the Church against the doctrines of Galileo and Darwin belongs. On the other hand, representatives of science have often made an attempt to arrive at fundamental judgments with respect to values and ends on the basis of scientific method, and in this way have set themselves in opposition to religion. These conflicts have all sprung from fatal errors.

Now, even though the realms of religion and science in themselves are clearly marked off from each other, nevertheless there exist between the two strong reciprocal relationships and dependencies. Though religion may be that which determines the goal, it has, nevertheless, learned from science, in the broadest sense, what means will contribute to the attainment of the goals it has set up. But science can only be created by those who are thoroughly imbued with the aspiration towards truth and understanding. This source of feeling, however, springs from the sphere of religion. To this there also belongs the faith in the possibility that the regulations valid for the world of existence are rational, that is, comprehensible to reason. I cannot conceive of a genuine scientist without that profound faith. The situation may be expressed by an image: Science without religion is lame, religion without science is blind.

Though I have asserted above that in truth a legitimate conflict between religion and science cannot exist I must nevertheless qualify this assertion once again on an essential point, with reference to the actual content of historical religions. This qualification has to do with the concept of God. During the youthful period of mankind's spiritual evolution human fantasy created gods in man's own image, who, by the operations of their will were supposed to determine, or at any rate to influence the phenomenal world. Man sought to alter the disposition of these gods in his own favor by means of magic and prayer. The idea of God in the religions taught at present is a sublimation of that old conception of the gods. Its anthropomorphic character is shown, for instance, by the fact that men appeal to the Divine Being in prayers and plead for the fulfilment of their wishes.

Nobody, certainly, will deny that the idea of the existence of an omnipotent, just and omnibeneficent personal God is able to accord man solace, help, and guidance; also, by virtue of its simplicity it is accessible to the most undeveloped mind. But, on the other hand, there are decisive weaknesses attached to this idea in itself, which have been painfully felt since the beginning of history. That is, if this being is omnipotent then every occurrence, including every human action, every human thought, and every human feeling and aspiration is also His work; how is it possible to think of holding men responsible for their deeds and thoughts before such an almighty Being? In giving out punishment and rewards He would to a certain extent be passing judgment on Himself. How can this be combined with the goodness and righteousness ascribed to Him?

The main source of the present-day conflicts between the spheres of religion and of science lies in this concept of a personal God. It is the aim of science to establish general rules which determine the reciprocal connection of objects and events in time and space. For these rules, or laws of nature, absolutely general validity is required—not proven. It is mainly a program, and faith in the possibility of its accomplishment in principle is only founded on partial successes. But hardly anyone could be found who would deny these partial successes and ascribe them to human self-deception. The fact that on the basis of such laws we are able to predict the temporal behavior of phenomena in certain domains with great precision and certainty is deeply embedded in the consciousness of the modern man, even though he may have grasped very little of the contents of those laws. He need only consider that planetary courses within the solar system may be calculated in advance with great exactitude on the basis of a limited number of simple laws. In a similar way, though not with the same precision, it is possible to calculate in advance the mode of operation of an electric motor, a transmission system, or of a wireless apparatus, even when dealing with a novel development.

To be sure, when the number of factors coming into play in a phenomenological complex is too large scientific method in most cases fails us. One need only think of the weather, in which case prediction even for a few days ahead is impossible. Nevertheless no one doubts that we are confronted with a causal connection whose causal components are in the main known to us. Occurrences in this domain are beyond the reach of exact prediction because of the variety of factors in operation, not because of any lack of order in nature.

We have penetrated far less deeply into the regularities obtaining within the realm of living things, but deeply enough nevertheless to sense at least the rule of fixed necessity. One need only think of the systematic order in heredity, and in the effect of poisons, as for instance alcohol, on the behavior of organic beings. What is still lacking here is a grasp of connections of profound generality, but not a knowledge of order in itself.

The more a man is imbued with the ordered regularity of all events the firmer becomes his conviction that there is no room left by the side of this ordered regularity for causes of a different nature. For him neither the rule of human nor the rule of divine will exists as an independent cause of natural events. To be sure, the doctrine of a personal God interfering with natural events could never be *refuted*, in the real sense, by science, for this doctrine can always take refuge in those domains in which scientific knowledge has not yet been able to set foot.

But I am persuaded that such behavior on the part of the representatives of religion would not only be unworthy but also fatal. For a doctrine which is able to maintain itself not in clear light but only in the dark, will of necessity lose its effect on mankind, with incalculable harm to human progress. In their struggle for the ethical good, teachers of religion must have the stature to give up the doctrine of a personal God, that is, give up that source of fear and hope which in the past placed such vast power in the hands of priests. In their labors they will have to avail themselves of those forces which are capable of cultivating the Good, the True, and the Beautiful in humanity itself. This is, to be sure, a more difficult but an incomparably more worthy task. After religious teachers accomplish the refining process indicated they will surely recognize with joy that true religion has been ennobled and made more profound by scientific knowledge.

If it is one of the goals of religion to liberate mankind as far as possible from the bondage of egocentric cravings, desires, and fears, scientific reasoning can aid religion in yet another sense. Although it is true that it is the goal of science to discover rules which permit the association and foretelling of facts, this is not its only aim. It also seeks to reduce the connections discovered to the smallest possible number of mutually independent conceptual elements. It is in this striving after the rational unification of the manifold that it encounters its greatest successes, even though it is precisely this attempt which causes it to run the greatest risk of falling a prey to illusions. But whoever has undergone the intense experience of successful advances made in this domain, is moved by profound reverence for the rationality made manifest in existence. By way of the understanding he achieves a far-reaching emancipation from the shackles of personal hopes and desires, and thereby attains that humble attitude of mind towards the grandeur of reason incarnate in existence, and which, in its profoundest depths, is inaccessible to man. This attitude, however, appears to me to be religious, in the highest sense of the word. And so it seems to me that science not only purifies the religious impulse of the dross of its anthropomorphism but also contributes to a religious spiritualization of our understanding of life.

The further the spiritual evolution of mankind advances,

the more certain it seems to me that the path to genuine religiosity does not lie through the fear of life, and the fear of death, and blind faith, but through striving after rational knowledge. In this sense I believe that the priest must become a teacher if he wishes to do justice to his lofty educational mission.

8

Werner Heisenberg,   PHYSICS AND PHILOSOPHY

James Jeans,   THE ASTRONOMICAL HORIZON

1. The uncertainty principle, first put forward by Heisenberg, seems to place a limit on the precision with which one can observe nature and thus on what one can know from nature. What does this limit imply about the existence of a knowable reality?

2. Quantum mechanics allows one only to calculate the probabilities of various occurences. Thus the actual outcome of an experiment at the atomic level is not knowable in advance. How does this compare to the known cause, predictable effect of Newtonian physics?

3. Newtonian physics was the basis for a world view known as the clockwork universe. When applied to the human mind, this even seemed to imply that all human thought was predetermined. What does quantum theory do to this view and how does it affect the notion of free will?

4. The so-called romantic movement in literature was in part a rebellion against the mechanistic world view of Newtonian physics. Do you see any current movements in society which might be in part rebellion against the probabilistic world view of Heisenberg?

5. How does one's perception of the universe affect his perception of self?

6. How does one explain what seems to be an inner drive within man to know about the universe in which he exists - and continually to explore the bounds of that universe?

7. Much of what you have read in this volume has dealt with the existence of an objective reality, outside of ourselves, which can be known. Are you comfortable with the idea that what we know is a construct of our minds and that there may be no such reality?

Werner Heisenberg (1901-   ) was awarded the Nobel Prize in 1932 in physics for his seminal work in developing the theory known as quantum mechanics. In addition to his work in theoretical physics, Heisenberg has written several exceptionally clear works for laymen, describing the advances which he pioneered in physics. He is certainly one of the most important twentieth-century physicists.

The History of Quantum Theory

THE origin of quantum theory is connected with a well-known phenomenon, which did not belong to the central parts of atomic physics. Any piece of matter when it is heated starts to glow, gets red hot and white hot at higher temperatures. The color does not depend much on the surface of the material, and for a black body it depends solely on the temperature. Therefore, the radiation emitted by such a black body at high temperatures is a suitable object for physical research; it is a simple phenomenon that should find a simple explanation in terms of the known laws for radiation and heat. The attempt made at the end of the nineteenth century by Lord Rayleigh and Jeans failed, however, and revealed serious difficulties. It would not be possible to describe these difficulties here in simple terms. It must be sufficient to state that the application of the known laws did not lead to sensible results. When Planck, in 1895, entered this line of research he tried to turn the problem from radiation to the radiating atom. This turning did not remove any of the difficulties inherent in the problem, but it simplified the interpretation of the empirical facts. It was just at this time, during the summer of 1900, that Curlbaum and Rubens in Berlin had made very accurate new measurements of the spectrum of heat radiation. When Planck heard of these results he tried to represent them by simple mathematical formulas which looked plausible from his research on the general connection between heat and radiation. One day Planck and Rubens met for tea in Planck's home and compared Rubens' latest results with a new formula suggested by Planck. The comparison showed a complete agreement. This was the discovery of Planck's law of heat radiation.

It was at the same time the beginning of intense theoretical work for Planck. What was the correct physical interpretation of the new formula? Since Planck could, from his earlier work, translate his formula easily into a statement about the radiating atom (the so-called oscillator), he must soon have found that his formula looked as if the oscillator could only contain discrete quanta of energy—a result that was so different from anything known in classical physics that he certainly must have refused to believe it in the beginning. But in a period of most intensive work during the summer of 1900 he finally convinced himself that there was no way of escaping from this conclusion. It was told by Planck's son that his father spoke to him about his new ideas on a long walk through the Grunewald, the wood in the suburbs of Berlin. On this walk he explained that he felt he had possibly made a discovery of the first rank, comparable perhaps only to the discoveries of Newton. So Planck must have realized at this time that his formula had touched the foundations of our description of nature, and that these foundations would one day start to move from their traditional present location toward a new and as yet unknown position of stability. Planck, who was conservative in his whole outlook, did not like this consequence at all, but he published his quantum hypothesis in December of 1900.

The idea that energy could be emitted or absorbed only in discrete energy quanta was so new that it could not be fitted into the traditional framework of physics. An attempt by Planck to reconcile his new hypothesis with the older laws of radiation failed in the essential points. It took five years until the next step could be made in the new direction.

This time it was the young Albert Einstein, a revolutionary genius among the physicists, who was not afraid to go further away from the old concepts. There were two problems in which he could make use of the new ideas. One was the so-called photoelectric effect, the emission of electrons from metals under the influence of light. The experiments, especially those of Lenard, had shown that the energy of the emitted electrons did not depend on the intensity of the light, but only on its color or, more precisely, on its frequency. This could not be understood on the basis of the traditional theory of radiation. Einstein could explain the observations by interpreting Planck's hypothesis as saying that light consists of quanta of energy traveling through space. The energy of one light quantum should, in agreement with Planck's assumptions, be equal to the frequency of the light multiplied by Planck's constant.

The other problem was the specific heat of solid bodies. The traditional theory led to values for the specific heat which fitted the observations at higher temperatures but disagreed with them at low ones. Again Einstein was able to show that one could understand this behavior by applying the quantum hypothesis to the elastic vibrations of the atoms in the solid body. These two results marked a very important advance, since they revealed the presence of Planck's quantum of action—as his constant is called among the physicists—in several phenomena, which had nothing immediately to do with heat radiation. They revealed at the same time the deeply revolutionary character of the new hypothesis, since the first of them led to a description of light completely different from the traditional wave picture. Light could either be interpreted as consisting of electromagnetic waves, according to Maxwell's theory, or as consisting of light quanta, energy packets traveling through space with high velocity. But could it be both? Einstein knew, of course, that the well-known phenomena of diffraction and interference can be explained only on the basis of the wave picture. He was not able to dispute the complete contradiction between this wave picture and the idea of the light quanta; nor did he even attempt to remove the inconsistency of this interpretation. He simply took the contradiction as something which would probably be understood only much later.

In the meantime the experiments of Becquerel, Curie and Rutherford had led to some clarification concerning the structure of the atom. In 1911 Rutherford's observations on the interaction of α-rays penetrating through matter resulted in his famous atomic model. The atom is pictured as consisting of a nucleus, which is positively charged and contains nearly the total mass of the atom, and electrons, which circle around the nucleus like the planets circle around the sun. The chemical bond between atoms of different elements is explained as an interaction between the outer electrons of the neighboring atoms; it has not directly to do with the atomic nucleus. The nucleus determines the chemical behavior of the atom through its charge which in turn fixes the number of electrons in the neutral atom. Initially this model of the atom could not explain the most characteristic feature of the atom, its enormous stability. No planetary system following the laws of Newton's mechanics would ever go back to its original configuration after a collision with another such system. But an atom of the element carbon, for instance, will still remain a carbon atom after any collision or interaction in chemical binding.

The explanation for this unusual stability was given by Bohr in 1913, through the application of Planck's quantum hypothesis. If the atom can change its energy only by discrete energy quanta, this must mean that the atom can exist only in discrete stationary states, the lowest of which is the normal state of the atom. Therefore, after any kind of interaction the atom will finally always fall back into its normal state.

By this application of quantum theory to the atomic model, Bohr could not only explain the stability of the atom but also, in some simple cases, give a theoretical interpretation of the line spectra emitted by the atoms after the excitation through electric discharge or heat. His theory rested upon a combination of classical mechanics for the motion of the electrons with quantum conditions, which were imposed upon the classical motions for defining the discrete stationary states of the system. A consistent mathematical formulation for those conditions was later given by Sommerfeld. Bohr was well aware of the fact that the quantum conditions spoil in some way the consistency of Newtonian mechanics. In the simple case of the hydrogen atom one could calculate from Bohr's theory the frequencies of the light emitted by the atom, and the agreement with the observations was perfect. Yet these frequencies were different from the orbital frequencies and their harmonics of the electrons circling around the nucleus, and this fact showed at once that the theory was still full of contradictions. But it contained an essential part of the truth. It did explain qualitatively the chemical behavior of the atoms and their line spectra; the existence of the discrete stationary states was verified by the experiments of Franck and Hertz, Stern and Gerlach.

Bohr's theory had opened up a new line of research. The great amount of experimental material collected by spectroscopy through several decades was now available for information about the strange quantum laws governing the motions of the electrons in the atom. The many experiments of chemistry could be used for the same purpose. It was from this time on that the physicists learned to ask the right questions; and asking the right question is frequently more than halfway to the solution of the problem.

What were these questions? Practically all of them had to do with the strange apparent contradictions between the results of different experiments. How could it be that the same radiation that produces interference patterns, and therefore must consist of waves, also produces the photoelectric effect, and therefore must consist of moving particles? How could it be that the frequency of the orbital motion of the electron in the atom does not show up in the frequency of the emitted radiation? Does this mean that there is no orbital motion? But if the idea of orbital motion should in incorrect, what happens to the electrons inside the atom? One can see the electrons move through a cloud chamber, and sometimes they are knocked out of an atom; why should they not also move within the atom? It is true that they might be at rest in the normal state of the atom, the state of lowest energy. But there are many states of higher energy, where the electronic shell has an angular momentum. There the electrons cannot possibly be at rest. One could add a number of similar examples. Again and again one found that the attempt to describe atomic events in the traditional terms of physics led to contradictions.

Gradually, during the early twenties, the physicists became accustomed to these difficulties, they acquired a certain vague knowledge about where trouble would occur, and they learned to avoid contradictions. They knew which description of an atomic event would be the correct one for the special experiment under discussion. This was not sufficient to form a consistent general picture of what happens in a quantum process, but it changed the minds of the physicists in such a way that they somehow got into the spirit of quantum theory. Therefore, even some time before one had a consistent formulation of quantum theory one knew more or less what would be the result of any experiment.

One frequently discussed what one called ideal experiments. Such experiments were designed to answer a very critical question irrespective of whether or not they could actually be carried out. Of course it was important that it should be possible in principle to carry out the experiment, but the technique might be extremely complicated. These ideal experiments could be very useful in clarifying certain problems. If there was no agreement among the physicists about the result of such an ideal experiment, it was frequently possible to find a similar but simpler experiment that could be carried out, so that the experimental answer contributed essentially to the clarification of quantum theory.

The strangest experience of those years was that the paradoxes of quantum theory did not disappear during this process of clarification; on the contrary, they became even more marked and more exciting. There was, for instance, the experiment of Compton on the scattering of X-rays. From earlier experiments on the interference of scattered light there could be no doubt that scattering takes place essentially in the following way: The incident light wave makes an electron in the beam vibrate in the frequency of the wave; the oscillating electron then emits a spherical wave with the same frequency and thereby produces the scattered light. However, Compton found in 1923 that the frequency of scattered X-rays was different from the frequency of the incident X-ray. This change of frequency could be formally understood by assuming that scattering is to be described as collision of a light quantum with an electron. The energy of the light quantum is changed during the collision; and since the frequency times Planck's constant should be the energy of the light quantum, the frequency also should be changed. But what happens in this interpretation of the light wave? The two experiments—one on the interference of scattered light and the other on the change of frequency of the scattered light—seemed to contradict each other without any possibility of compromise.

By this time many physicists were convinced that these apparent contradictions belonged to the intrinsic structure of atomic physics. Therefore, in 1924 de Broglie in France tried to extend the dualism between wave description and particle description to the elementary particles of matter, primarily to the electrons. He showed that a certain matter wave could "correspond" to a moving electron, just as a light wave corresponds to a moving light quantum. It was not clear at the time what the word "correspond" meant in this connection. But de Broglie suggested that the quantum condition in Bohr's theory should be interpreted as a statement about the matter waves. A wave circling around a nucleus can for geometrical reasons only be a stationary wave; and the perimeter of the orbit must be an integer multiple of the wave length. In this way de Broglie's idea connected the quantum condition, which always had been a foreign element in the mechanics of the electrons, with the dualism between waves and particles.

In Bohr's theory the discrepancy between the calculated orbital frequency of the electrons and the frequency of the emitted radiation had to be interpreted as a limitation to the concept of the electronic orbit. This concept had been somewhat doubtful from the beginning. For the higher orbits, however, the electrons should move at a large distance from the nucleus just as they do when one sees them moving through a cloud chamber. There one should speak about electronic orbits. It was therefore very satisfactory that for these higher orbits the frequencies of the emitted radiation approach the orbital frequency and its higher harmonics. Also Bohr had already suggested in his early papers that the intensities of the emitted spectral lines approach the intensities of the corresponding harmonics. This principle of correspondence had proved very useful for the approximative calculation of the intensities of spectral lines. In this way one had the impression that Bohr's theory gave a qualitative but not a quantitative description of what happens inside the atom; that some new feature of the behavior of matter was qualitatively expressed by the quantum conditions, which in turn were connected with the dualism between waves and particles.

The precise mathematical formulation of quantum theory finally emerged from two different developments. The one started from Bohr's principle of correspondence. One had to give up the concept of the electronic orbit but still had to maintain it in the limit of high quantum numbers, i.e., for the large orbits. In this latter case the emitted radiation, by means of its frequencies and intensities, gives a picture of the electronic orbit; it represents what the mathematicians call a Fourier expansion of the orbit. The idea suggested itself that one should write down the mechanical laws not as equations for the positions and velocities of the electrons but as equations for the frequencies and amplitudes of their Fourier expansion. Starting from such equations and changing them very little one could hope to come to relations for those quantities which correspond to the frequencies and intensities of the emitted radiation, even for the small orbits and the ground state of the atom. This plan could actually be carried out; in the summer of 1925 it led to a mathematical formalism called matrix mechanics or, more generally, quantum mechanics. The equations of motion in Newtonian mechanics were replaced by similar equations between matrices; it was a strange experience to find that many of the old results of Newtonian mechanics, like conservation of energy, etc., could be derived also in the new scheme. Later the investigations of Born, Jordan and Dirac showed that the matrices representing position and momentum of the electron do not commute. This latter fact demonstrated clearly the essential difference between quantum mechanics and classical mechanics.

The other development followed de Broglie's idea of matter waves. Schrödinger tried to set up a wave equation for de Broglie's stationary waves around the nucleus. Early in 1926 he succeeded in deriving the energy values of the stationary states of the hydrogen atom as "Eigenvalues" of his wave equation and could give a more general prescription for transforming a given set of classical equations of motion into a corresponding wave equation in a space of many dimensions. Later he was able to prove that his formalism of wave mechanics was mathematically equivalent to the earlier formalism of quantum mechanics.

Thus one finally had a consistent mathematical formalism,

which could be defined in two equivalent ways starting either from relations between matrices or from wave equations. This formalism gave the correct energy values for the hydrogen atom; it took less than one year to show that it was also successful for the helium atom and the more complicated problems of the heavier atoms. But in what sense did the new formalism describe the atom? The paradoxes of the dualism between wave picture and particle picture were not solved; they were hidden somehow in the mathematical scheme.

A first and very interesting step toward a real understanding of quantum theory was taken by Bohr, Kramers and Slater in 1924. These authors tried to solve the apparent contradiction between the wave picture and the particle picture by the concept of the probability wave. The electromagnetic waves were interpreted not as "real" waves but as probability waves, the intensity of which determines in every point the probability for the absorption (or induced emission) of a light quantum by an atom at this point. This idea led to the conclusion that the laws of conservation of energy and momentum need not be true for the single event, that they are only statistical laws and are true only in the statistical average. This conclusion was not correct, however, and the connections between the wave aspect and the particle aspect of radiation were still more complicated.

But the paper of Bohr, Kramers and Slater revealed one essential feature of the correct interpretation of quantum theory. This concept of the probability wave was something entirely new in theoretical physics since Newton. Probability in mathematics or in statistical mechanics means a statement about our degree of knowledge of the actual situation. In throwing dice we do not know the fine details of the motion of our hands which determine the fall of the dice and therefore we say that the probability for throwing a special number is just one in six. The probability wave of Bohr, Kramers, Slater, however, meant more than that; it meant a tendency for something. It was a quantitative version of the old concept of "potentia" in Aristotelian philosophy. It introduced something standing in the middle between the idea of an event and the actual event, a strange kind of physical reality just in the middle between possibility and reality.

Later when the mathematical framework of quantum theory was fixed, Born took up this idea of the probability wave and gave a clear definition of the mathematical quantity in the formalism, which was to be interpreted as the probability wave. It was not a three-dimensional wave like elastic or radio waves, but a wave in the many-dimensional configuration space, and therefore a rather abstract mathematical quantity.

Even at this time, in the summer of 1926, it was not clear in every case how the mathematical formalism should be used to describe a given experimental situation. One knew how to describe the stationary states of an atom, but one did not know how to describe a much simpler event—as for instance an electron moving through a cloud chamber.

When Schrödinger in that summer had shown that his formalism of wave mechanics was mathematically equivalent to quantum mechanics he tried for some time to abandon the idea of quanta and "quantum jumps" altogether and to replace the electrons in the atoms simply by his three-dimensional matter waves. He was inspired to this attempt by his result, that the energy levels of the hydrogen atom in his theory seemed to be simply the eigenfrequencies of the stationary matter waves. Therefore, he thought it was a mistake to call them energies; they were just frequencies. But in the discussions which took place in the autumn of 1926 in Copenhagen between Bohr and Schrödinger and the Copenhagen group of physicists it soon became apparent that such an interpretation would not even be sufficient to explain Planck's formula of heat radiation.

During the months following these discussions an intensive study of all questions concerning the interpretation of quantum theory in Copenhagen finally led to a complete and, as many physicists believe, satisfactory clarification of the situation. But it was not a solution which one could easily accept. I remember discussions with Bohr which went through many hours till very late at night and ended almost in despair; and when at the end of the discussion I went alone for a walk in the neighboring park I repeated to myself again and again the question: Can nature possibly be as absurd as it seemed to us in these atomic experiments?

The final solution was approached in two different ways. The one was a turning around of the question. Instead of asking: How can one in the known mathematical scheme express a given experimental situation? the other question was put: Is it true, perhaps, that only such experimental situations can arise in nature as can be expressed in the mathematical formalism? The assumption that this was actually true led to limitations in the use of those concepts that had been the basis of classical physics since Newton. One could speak of the position and of the velocity of an electron as in Newtonian mechanics and one could observe and measure these quantities. But one could not fix both quantities simultaneously with an arbitrarily high accuracy. Actually the product of these two inaccuracies turned out to be not less than Planck's constant divided by the mass of the particle. Similar relations could be formulated for other experimental situations. They are usually called relations of uncertainty or principle of indeterminacy. One had learned that the old concepts fit nature only inaccurately.

The other way of approach was Bohr's concept of complementarity. Schrödinger had described the atom as a system not of a nucleus and electrons but of a nucleus and matter waves. This picture of the matter waves certainly also contained an element of truth. Bohr considered the two pictures—particle picture and wave picture—as two complementary descriptions of the same reality. Any of these descriptions can be only partially true, there must be limitations to the use of the particle concept as well as of the wave concept, else one could not avoid contradictions. If one takes into account those limitations which can be expressed by the uncertainty relations, the contradictions disappear.

In this way since the spring of 1927 one has had a consistent interpretation of quantum theory, which is frequently called the "Copenhagen interpretation." This interpretation received its crucial test in the autumn of 1927 at the Solvay conference in Brussels. Those experiments which had always led to the worst paradoxes were again and again discussed in all details, especially by Einstein. New ideal experiments were invented to trace any possible inconsistency of the theory, but the theory was shown to be consistent and seemed to fit the experiments as far as one could see.

The details of this Copenhagen interpretation will be the subject of the next chapter. It should be emphasized at this point that it has taken more than a quarter of a century to get from the first idea of the existence of energy quanta to a real understanding of the quantum theoretical laws. This indicates the great change that had to take place in the fundamental concepts concerning reality before one could understand the new situation.

The British mathematician, James Jeans (1877-1946) has emerged as one of the premier cosmologists of the twentieth-century. In addition to major works in sound, electricity, and radiation, Jeans provided an exceptionally articulate view of the ever-changing cosmos. He not only did significant original research but also described that research, and the work of others, in a way which won for him enormous popular acclaim.

THE ASTRONOMICAL HORIZON

THE history of astronomy, it has been truly said, is a history of receding horizons. Every increase in telescopic power and—what is hardly less important—every gain in the sensitivity of photographic plates, has opened up new depths of space for exploration, and the centre of astronomical interest has tended to follow the limits of vision. Until the middle of the last century, interest was centred in the planets of the solar system; after that it shifted to the stars of the galactic system; now many astronomers find their main interest in the remote nebulae which lie at the extreme limits of vision of our largest telescopes—the astronomical horizon. Each of these advances—from planets to stars, and from stars to nebulae—has involved a vast march forward into space, for the nearest stars are a million times as distant as the nearest planets, and the nebulae are nearly a million times as distant as the stars.

Knowledge too has its horizons. Every science contains a hard core of well-established fact and, surrounding this, a dim nebulous region in which it is hard to distinguish truth from error, a region of conjecture and doubtful hypotheses. Here is the frontier of knowledge, the region in which progress is being made. In astronomy it is closely associated with the frontier of known space. It is to these vague regions that I shall venture to take you this afternoon, approaching them through regions of well ascertained fact, and pausing here and there to recollect how this nucleus of firm knowledge was gained—for the methods by which the frontier was reached are still of service when we try to make progress beyond.

The ancients thought of the universe as quite a small affair, with our planet as its main ingredient, and the sky as its not very remote ceiling. Then, about 2,200 years ago, Aristarchus of Samos, a teacher in Alexandria, published a book *On the Sizes and Distances of the Sun and Moon*. Astronomers had made observations for many thousands of years before Aristarchus, but he was the first to discuss his observations in a true mathematical spirit. The observations were far from accurate, but he deduced from them by strict reasoning that the sun must be many thousands of times as big as the earth. Then, probably seeing the absurdity of supposing so massive a sun to revolve round so much smaller an earth, he introduced the hypotheses (1) 'that the fixed stars and the sun remain motionless, that the earth revolves about the sun, ... and (2) that the sphere of the fixed stars is so great' that the Earth's orbit 'bears the same proportion to the distance of the fixed stars as the centre of the sphere bears to its surface'. In brief, he reduces the earth's orbit to a point.

Here is the first known clear statement that the earth revolves round the sun, and also one of the earliest known statements that the stars are immensely distant. The two statements are of course not independent; the latter is a direct consequence of the former. For if the earth revolved round the sun, and the stars were not very distant, they would appear to change their relative positions as the earth moved towards some and away from others in its journey round the sun. Aristarchus, understanding this and noticing no changes, saw that the world of stars must be very great.

His thumb-nail sketch of the universe contained the essence of the truth, but just failed to start astronomy on the right road. The authority of Aristotle, of Plato, and of many others stood against him, with the result that for nearly 2,000 years sound views as to the structure of the universe were driven into hiding. In the second century after Christ we find the Alexandrian astronomer Ptolemy marshalling arguments to prove that the earth could not move in space. Some were quite absurd, as, for instance that a motion of the earth through space would create such a wind that birds who ventured out into it would never be able to get back to their nests. He repeated the argument that if the earth were moving the stars would appear to change their relative positions, but did not mention Aristarchus' solution that the stars may be too distant for the changes to be noticeable.

In a later age the authority of Aristotle was first reinforced and then superseded by the authority of the Christian Church, but from the Church also came the most damaging criticisms of the current doctrines. From the fourteenth century on we find a succession of ecclesiastics—Oresme Bishop of Lisieux, Cardinal Nicolas of Cusa, Canon Copernicus of Frauenburg, and the monk Giordano Bruno—all declaring against the Aristotelian cosmology and maintaining that the earth moves, that the stars are other worlds, and that the earth's orbit in space is infinitesimal in comparison with the universe of stars—'as a point is to a clod' was the comparison Copernicus used when he explained, as Aristarchus had done before him, why we notice no changes in the positions of the stars. Bruno went further, maintaining that the infinite goodness of God required that the stars should be infinite in number, and arguing that, as an infinity can have no centre, it is impossible for either the sun or the earth to stand at the centre of the universe. His doctrines probably affected human thought even more profoundly than those of Copernicus, and certainly contained more potential danger to the tenets of orthodoxy. He fell into the hands of the Inquisition, and was condemned and burned at the stake in 1600.

Efforts to explain why the stars showed no changes of position ended in the discovery that there was nothing to explain. In 1718 Halley found that Sirius and Arcturus had moved substantially since Ptolemy had charted their positions some sixteen centuries earlier. From now on stellar positions and their changes were studied systematically, and in time yearly oscillations were found which obviously resulted from the earth's motion round the sun. The more distant a star, the more minute this yearly oscillation would be, so that a measurement of its amount would fix the distance of the star—in terms of course of the dimensions of the earth's orbit. About 106 years ago the distances of three of the nearer stars were measured by this method. Astronomy had now become possessed of a measuring-rod for stellar distances, and only technical difficulties stood in the way of a complete survey of the skies.

Long before this, however, it had become recognized that the stars were suns like our own, so that their comparative faintness showed that they must be very distant. If every star emitted exactly the same amount of light as the sun, a star's faintness would of course provide a precise measure of its distance; Newton had estimated stellar distances in this way, with results that have proved to be not too far from the truth. This method can no longer be used, for we know that its basic supposition is false; the stars shine with very unequal lights, some emitting less than a 300,000th part of the light of the sun, and some more than 300,000 times as much—a range of 90 thousand million to one in luminosity. But the method is still very useful in a modified form. There are special classes of stars—as for instance certain variable stars of assigned periods (Cepheids)—in which all members of any one class are found to be of the same luminosity. These stars provide a system of standard beacons scattered through space, and by their relative faintness and brightness, as seen by us, enable us to measure stellar distances with some accuracy. But even these methods fail for the most distant stars of all, those which form the star-dust of the Milky Way. And these are just the stars that matter when we try to plot out the dimensions and shape of our system of stars. For this problem other methods are needed, and have been found.

Plate I consists of four photographs of the same small area of the sky, an area in which no stars are visible to the naked eye. They show what we should see if we examined this bit of the sky visually through telescopes of apertures 4, 16, 64, and about 160 inches respectively. If space were perfectly transparent to light, pictures (2) and (3) would each probe just four times as far into space as its predecessor, and so would disclose 64 times as big a volume of space. Thus if the stars were uniformly scattered through infinite space, each picture would show 64 times as many stars as the picture before. But a simple count will show that the ratio of increase is nothing like as great as 64, and the same applies, *mutatis mutandis*, to the step from the third picture to the last. I₁, then, space is perfectly transparent to light, we must conclude that the stars are not uniformly scattered through infinite space, and that the later photographs already penetrate into regions in which the stars have begun to thin out.

Sir William Herschel reached similar conclusions when he used a primitive form of this method. He noted the distances at which the thinning seemed to begin in different parts of the sky, and concluded that the system of stars was shaped like a flat disk, with the sun at, or very near to, its centre.

This description survived until quite recently, but it is unsound. It was based on the supposition that space is perfectly transparent to light, and recent advances in astronomical knowledge and photographic technique have made any such supposition quite untenable. When we examine the sky visually or by a short exposure photograph, we see little beyond stars which appear as mere points of light, as on Plate I. But a long exposure reveals something very different. Plate II shows a long exposure photograph of the Pleiades, and here the stars no longer appear as points of light, but as extensive masses of what we call nebulosity. This and other methods of study have shown that the whole of interstellar space is occupied by a sort of fog of obscuring matter. Sometimes, as in Plate II, the fog may be lighted up by the stars within—much as a street lamp may light up the fog in its vicinity, although the physical process is very different. Perhaps, as in Plate VII (facing p. 18), we may find thin, filmy wisps of nebulosity which extend over a whole constellation, and are again raised to luminescence by the hot stars within. Or, as in Plate VIII (facing p. 19), the obscuring matter may show itself in its true character, blotting out the light from the stars beyond and so producing an impression of black patches, or even of holes, in the sky.

This obscuring fog prevents our seeing more than a certain distance through space; it imposes a definite range of visibility. When we walk in a forest on a foggy day, we see only those trees which are within a certain distance of us, so that we are at the centre of all the trees we can see. But we must not suppose that we are at the centre of the forest; we are merely at the centre of our individual sphere of visibility. It is the same with the stars.

Nevertheless Herschel, finding that we were at the centre of all the stars we can see, concluded that we were at the centre of the whole system of stars, and in this he was followed by Kapteyn as late as 1922. But again we are only at the centre of our sphere of visibility, the interstellar fog preventing our seeing the more distant stars of our system. Thus the discovery of the dimensions and shape of this system again· calls for new methods, and again they have been found.

Although the light from a single star cannot penetrate far through the fog, there are other objects in space—the 'globular clusters', close groups of millions of stars—which are brilliant enough to penetrate through almost the whole of the fog. They contain standard stars of known luminosity, so that we can measure their distances and plot out their arrangement. A study by Shapley showed that they occupy a disk-shaped volume of space, somewhat more than 100,000 light-years in radius. It was Herschel's disk-shaped figure repeated on a larger scale, with the difference that the sun did not stand at its centre but about 40,000 light-years away from the centre. It began to seem likely that this figure, rather than Herschel's, must map out the confines of the galactic system of stars.

This conjecture subsequently received confirmation in an unexpected way. A detailed study of the motions of the stars shows that they do not move at random, but after a plan— roughly they move like the planets round the sun. In the solar system the inner planets move faster round the sun than the outer, and so continually overtake the latter in their orbits. The stars of the galactic system are found to move in a similar way, and the centre round which they move proves to be the centre of Shapley's disk, those which are nearest to this centre moving fastest. The sun takes about 250 million years to complete a revolution round this centre and, as the radius of its orbit is about 40,000 light-years, it must move at about 270 km. a second. Actually the situation is not quite so simple as this, since the stars have small random motions of their own superposed on to their regular orbital motions, but these need not concern us here.

We know that the planets would scatter into space if it were not for the restraining force of the sun's gravitational pull. It is the same with the stars; they would scatter into space if it were not for the gravitational pull of the galaxy as a whole. We can estimate the amount of this pull from the particulars of the sun's orbit just given, and so can calculate the total mass of the galaxy from which the pull originates. It is found that this must be of the order of

150,000 million suns like our own. The largest telescope in the world shows only about 1,500 million stars, which thus represent less than one per cent. of the total mass; the remaining 99 per cent. is invisible to us. It will consist of stars which are too distant to be seen through the interstellar fog, stars which are too dark or too faint to be seen anyhow, and the matter of the fog itself.

Large though this mass is, it occupies so great a space that stars are rare occurrences in it. Let us attempt to construct a scale model, taking a large hall as the space available for the galaxy. Let us smoke in it until it contains 150,000 million particles of tobacco smoke, to achieve which we need only smoke a millionth of a gram of tobacco. We now have the right number of particles to represent the stars, but what about their size? Andrade and Parker find that the average diameter of a particle of tobacco smoke is rather less than a millionth of an inch, whence it appears that the particles are far too big—millions of times too big—to represent the stars. Even a single atom is too big to represent a star; its radius represents that of the orbit of Pluto—40 times the radius of the orbit of the earth, 8,000 times the radius of the sun. Thus after we have smoked our millionth of a gram of tobacco, the hall will be millions of times more crowded with tobacco particles than space is with stars.

Galileo's first telescope (1609) had shown nothing more distant than such stars. Three years later Marius observed an object of different type which looked 'like a candle-light seen through horn'. He had discovered the first of the great extra-galactic nebulae, that which we now call the Great Nebula in Andromeda (Plate III, facing p. 8).

Millions of these objects are now known, the majority being of characteristic and regular shapes. As far back as 1755 Kant suggested that they were huge systems of stars, probably very similar to our own galactic system. Herschel too conjectured that sufficient telescopic power would resolve them into clusters of stars. The needed power was not forthcoming until 1924 when Edwin Hubble, working with the great 100-inch telescope at Mount Wilson, succeeded in resolving some of the outer parts of the Andromeda nebula into separate stars (Plate IV, opposite). The final resolution of the central parts of this nebula into stars was only effected a few months ago by Baade, also working at Mount Wilson. Many other nebulae also have been studied in detail, and parts of them resolved into stars; they differ greatly in size and shape, but all appear to be stellar systems similar to our own galactic system.

Usually they are found to contain a profusion of the variable stars and other standard objects that are used for measuring distances in our own system. This makes it possible to measure the distances of the nebulae, and also their sizes. The Great Nebula in Andromeda proves to be the nearest of all, at a distance of about 680,000 light-years; its dimensions are found to be comparable with those of the galactic system.

The nebulae are of many different shapes, but it is found to be a general law that nebulae of the same shape are also of the same size (approximately) and also of the same luminosity. Once again, then, nature has provided the astronomer with standard articles—this time, the nebulae themselves. Differences of apparent size and brightness can originate only in differences of distance, so that we can estimate the distances of nebulae which are far too remote for standard stars to be recognized in them—indeed, up to distances of about 500 million light-years, beyond which it is impossible to recognize the nebulae as such. As we pass outwards in space we find no thinning out of the nebulae such as Herschel found with the stars; right up to the limits of our vision, the number visible within any distance remains proportional to the cube of the distance. They are found to be fairly uniformly spaced, at an average distance of about two million light-years. To construct a scale model, we may think of dinner plates scattered through space at an average distance of about twenty feet, and orientated at random. We choose dinner plates to represent the nebulae because most, although by no means all, of them are of the same flat disk shape as our own galaxy. In Plate V (facing p. 16), the lower picture is that of such a nebula seen 'full face', while the upper is that of a probably very similar nebula seen 'edge on'. In the latter the dark band round the equator obviously represents a band of obscuring fog which shuts off the light of the stars inside.

Although on the whole the nebulae are scattered uniformly through space, yet here and there they are found to fall into groups and clusters. On Plate VI (facing p. 17), the upper picture shows a close group of three in the constellation of Pegasus, while the lower shows a much larger cluster in Coma Berenicis which is at a distance of about 50 million light-years from us. In each of these clusters, the members are all at approximately equal distances from us, and it is of interest to notice that all appear of approximately the same sizes and brightnesses.

Where the nebulae fall into clusters we can estimate their masses or, more briefly, weigh them. For a cluster can only be kept in being by the gravitational attractions of its constituent nebulae. The speeds of motion of the various members can be measured spectroscopically, and from these we can deduce the gravitational pull needed to keep the cluster from disintegrating. This in turn tells us the mass of the whole cluster, and hence, by a simple division, the average mass of all the nebulae of the cluster. Generally this comes out to be of the order of 100,000 million to 200,000 million suns, providing further evidence that these external galaxies are similar systems to our own.

If this were all, the description of the Universe would be very simple—vast aggregations of stars, containing about 150,000 million each, occupying space fairly uniformly as far as our telescopes can probe and we know not how far beyond. Such a description would, we may imagine, have proved very satisfying to an astronomer of a generation ago; he would probably have conjectured that this Chapter of the Book of Nature was closed, except perhaps for elaboration in points of detail and possible explorations into still farther depths of space; little would he have been able to imagine the strange, even fantastic, world he was just about to enter.

The door to the new world was the relativity theory of gravitation which Einstein propounded in 1915. Newton had taught that gravitational acceleration was caused by a force that followed the simple mathematical law of the inverse square of the distance. Einstein found that in his new world of relativity such a law became meaningless, so that Newton's force of gravitation was physically impossible.

He regained consistency by attributing acceleration, not to force but to a curvature of space, or, to be more precise, to a curvature of that four-dimensional blend of space and time which forms the background to the whole theory of relativity. The path of a planet or a cricket ball was no longer a curved path in a straight space, but a straight path in a curved space; the curvature was transferred from the path in which the projectile moved to the space in which it moved. This curvature was impressed on any region of space by matter in its proximity. The sun, for instance, crumples up the regions of space in which the planets perform their orbits.

This was found to give a perfect account of the motions of the planets round the sun, but the discussion of the universe as a whole raised more difficult problems. The whole of space contains matter, and the amount of matter in any region of space determines the curvature in that region. With exactly the right amount of matter the total curvature in the whole of space might be just enough to close it up into a finite universe, which would then be in equilibrium. In this case, space would not be like the finite space inside a sphere, but like the finite surface of a sphere, not like the interior of the Earth, but like its surface—finite but unbounded. For a specified density of matter in space, there would be only one size of space for which equilibrium would be possible.

Einstein supposed that astronomical space must be of this kind. Since the average masses and distribution of the nebulae were known from observation, it was possible to calculate what the size and total mass of the universe must be. The particular calculations that were made at this time do not matter much, for they are now superseded; more modern calculations suggest that, in a general way, there must be about as many nebulae in space as there are stars in each nebula—something of the order of 100,000 million —only about 10 million of which are visible in our telescopes. We may visualize the total number in the whole of space as equal to the number of raindrops in a heavy shower—perhaps one giving a tenth of an inch rainfall over the City of Oxford.

This particular picture of a space held in equilibrium by the matter it contained was shattered a few years later when

359

Friedmann and Lemaître showed that the arrangement it portrayed could not be permanent. Space was represented as a sort of coiled spring, its curvature being impressed on it by the matter it contained. If the matter in a particular region of space became less dense, this part of the space would tend to uncoil; if matter moved from one region to another, the curvature would change in both, and the universe would no longer be in equilibrium. The new forces thus brought into play might tend either to restore the original equilibrium or to accentuate the existing disequilibrium. Friedmann and Lemaître showed that they would do the latter. Thus the Einstein arrangement proved to be unstable, so that space left to itself would necessarily start either to expand or to contract. Before the mathematicians had decided which would actually happen, the observers at Mount Wilson announced that, to all appearances, space was actually expanding, and this at no small speed.

Astronomers cannot, of course, observe an expansion of space directly; they can only examine rays of light. In this case they had examined the spectra of the light emitted by distant nebulae and found a systematic displacement towards the red end of the spectrum, a phenomenon which usually indicates that the source of the light is increasing its distance from us, so that fewer waves of light arrive per second. The most distant nebulae of which the spectra can be measured with accuracy show displacements such as would be produced by recession at the terrific speed of about 26,000 miles a second, which is about one-seventh of the speed of light.

The amount of the displacement was found to vary from one nebula to another, being approximately proportional to the distance of the nebula. If, however, a cluster of nebulae was treated as a single unit, so that any random motions of individual nebulae were averaged out, then the displacement for the cluster as a whole was found to be almost exactly proportional to the distance of the cluster. This relation is obeyed so exactly that it obviously must represent something fundamental in the structure of the universe. The simplest interpretation is that it results from a uniform expansion of the whole of space consequent upon the instability of the

Einstein configuration, the moving nebulae showing currents in space much as moving particles of dust show currents in air. Yet we may wonder whether an expansion produced in this way would result in so exact a relation between speed and distance as is actually observed. Matter is not distributed in space with perfect uniformity, so that we could hardly expect the expansion resulting from instability to be so perfectly uniform as, on this interpretation, it appears to be. It is at least conceivable that there may be some deeper underlying reason for the law.

There has been a good deal of discussion as to whether these apparent nebular motions really occur, or whether we are interpreting something quite different as motion. It is the kind of question that would have seemed outstandingly important to our scientific ancestors of a generation ago, and its answer would have been fundamental in their description of the universe. To-day, at least if we accept the doctrines of relativity, it seems to mean very little. For what do we mean if we assert that the apparent motions are illusory? We cannot mean that the nebulae are standing still in absolute space; this is meaningless, since there is no absolute space. Do we then mean that the distance between any pair of nebulae remains unaltered with the passage of time? This again is meaningless. We cannot construct an arch of yard-sticks in space from nebula to nebula, and stand by to see whether it breaks up or not—and, if it did, we should still not know whether the universe was expanding or the yard-sticks contracting. We might measure the angle which one nebula subtended at another, but even if this remained always the same, we could not be sure that the universe and the nebula were not expanding together. In whatever way we try to explain what absence of motion means, there is always some snag to invalidate our explanation; indeed, if the theory of relativity is true, there must be.

The best we can do is to devise an explanation which will 'save the phenomena' without making any claim to absolute truth. That of a uniformly expanding universe is certainly one of the simplest, although it may conceivably be open to the objection mentioned above. Another is obtained by supposing that, instead of the universe continually increasing in size, its atoms are continually diminishing in size. When we look at a cluster of nebulae which is 50 million light-years distant, such as that shown in Plate VI (facing p. 17), we see it by light which left the nebula 50 million years ago, and so originated in the atoms of 50 million years back in time. We may imagine that these atoms were larger than the corresponding atoms of to-day, and so emitted light of longer wave-length; this explains at once why the observed spectral displacement is so exactly proportional to distance. This last explanation admits of many variants—as, for instance, that the electric attractions and repulsions between charged particles vary with the passage of time; this requires neither the universe nor the atom to change their dimensions. Another and very different description assumes that as a light-quantum ages during its long journey through space, its energy steadily and uniformly decreases; quantum-theory shows that this would result in the spectral lines being displaced by an amount proportional to the distance the light had travelled, which is precisely the relation observed for the light from the nebulae.

These attempts at pictorial descriptions provide us with a first glimpse of the strange new world through which astronomers and mathematicians have been trying to grope their way during the last quarter-century, and to which we shall return in a moment.

We have seen how Einstein's original theory provided a relation between the density of matter in space (which can be estimated from the observational data I have already mentioned) and the size of the universe. This assumed the universe to be at rest. If we picture it as expanding, Einstein's relation must, as Lemaître showed, be replaced by another, which connects the speeds of recession of the nebulae with the radius of the universe, and so—indirectly—with the amount of matter it contains. When the average density of matter and the speeds of recession are assigned, only one value is possible for the total mass of the universe. Using the best observational data available we find that this mass is about that of 10^{79} hydrogen atoms. Or we may avoid any reference to the special substance hydrogen by saying that the universe must consist of about 10^{79} protons and an equal number of electrons. If we had inserted other astronomical data, this number might have come out as 10^{78} or 10^{80}.

Whatever data we assume, the number comes out conspicuously large, and we wonder why the substance of the universe should be broken up into so many separate particles. We wonder, too, what the ultimate interpretation of this number may be—why this particular number rather than any other?

The astronomer of a generation ago might have attempted an answer in terms of a very anthropomorphic Creator manufacturing this number of particles, setting them free in space, and resting on the seventh day because the number seemed good to Him: the number 10^{79} would represent a choice on the part of the Creator.

In recent years Sir Arthur Eddington suggested a much more interesting answer. He believed that the number did not represent any sort of choice, but an inevitable necessity; he thought it was a pure mathematical constant of the same kind as e or π; something inherent in the nature of things, which could be discovered by the human intellect working alone—not, it is true, without utilizing any observational knowledge of the world of physics, but without calling upon any observation of the outer world of astronomy. He believed he had shown that its value must be precisely $\frac{3}{2} \times 136 \times 2^{256}$, of which the value ($1 \cdot 16 \times 10^{79}$) is certainly of the order suggested by telescopic observation of the astronomical universe.

I have said that this number is not a matter of choice, yet according to Eddington a small element of choice does enter into it. The choice is not on the part of the Creator, but of ourselves. Eddington's contention, in his own words, is 'not that there are N particles in the universe [N denoting the number mentioned above], but that anyone that accepts certain elementary principles of measurement must, if he is consistent, think that there are'. In another place he says that 'when we find the physicist actually does think there are N particles, . . . this is not because of any peculiarity in the external world'. In brief, this N-ness of the external world appears as a sort of Kantian 'category', a contribution of our own minds, not to the universe but to our interpretation of it.

Eddington's primary assumption is that the universe consists of particles of such a primitive kind that they can have no attributes beyond existence and non-existence; in assuming this he elects, as a matter of choice, to describe the universe in terms of what the physicist calls a 'particle picture'. Until about 1925 the physicist took it for granted that the universe was made up of particles, because he found that this assumption provided the simplest interpretation of the then known phenomena. Actually in depicting matter in this way he was, as we now know, making a choice, and he had to take the consequences of this choice; one of them, as Planck and Bohr found, was that he had to eject continuity from his scheme of nature. He now finds it more convenient for many purposes to picture matter as waves—waves which are too complex to be represented in space and time, and are too continuous to admit of division into discrete particles or other physical units, waves which are mathematical rather than physical, mental constructs rather than physical entities. The view that the universe consists of particles is only one of many possible views, and one of the consequences to which it commits us, according to Eddington, is that the particles must be N in number. The view that it consists of some other number N' of particles is not a permissible view, since it involves a logical inconsistency.

It is difficult to discuss all this, since Eddington did not claim to be able to give a rigorous proof of it, either logical or mathematical, but only adduced certain considerations which seemed to him to point to its validity, and these I personally find wholly unconvincing. But whether he is right or wrong, he can claim to have revealed to us a suggestive, fascinating and inspiring vision. Even if events should prove his train of thought to be entirely fallacious, the number N is still of great interest. It is no stranger to mathematical thought, especially in the form of its square root of which the value is $\sqrt{N} = 3.4 \times 10^{39}$. Mathematical numbers of precisely this order of magnitude have been noticed before; let me mention some.

In ordinary life, our unit of force is the weight of a pound or of a gram, but these units are a matter of human convention, or at best are based on accidental properties of the planet on which we happen to live. But purely observational physics can supply a unit of force which is independent of human conventions, and would be as suitable for use on Mars or on Sirius as it is here, namely, the electrical attraction between the electron and the proton in the hydrogen atom. Astronomy can also provide a universal unit of force—the gravitational attraction between the same two particles. It has often been remarked how unequal these two forces are; their ratio is about $2 \cdot 3 \times 10^{39}$. This ratio, let us notice, is a pure number which must somehow be inherent in the scheme of things—again like e or π. And its value is suspiciously near to Eddington's $\sqrt{N}$.

It is the same with the measurement of length. We terrestrials measure in feet or centimetres, which depend on the human body or the dimensions of our planet, but there is no need to be so parochial. Physics provides a natural unit of length, the so-called 'radius' of the electron, which is about 2×10^{-13} cm., while astronomy provides another—the radius of a universe which would contain all the matter of the present universe in equilibrium, or, if we like to adopt a more precise picture, the radius of the universe before expansion began, which, according to Eddington, was about 1,068 million light-years. Again the two units are grossly unequal, their ratio being about $5 \cdot 0 \times 10^{39}$.

It is the same with time. We employ the terrestrial day and year as units, but nature has provided a unit of cosmical significance in the recession of the nebulae. Since their speeds of recession are proportional to their distances, it follows that if the speed of recession of each nebula could be reversed, and the nebula made to move backwards with this reversed speed, all the nebulae would finally meet in some quite small region of space. On inserting the observed values for the speeds of the nebulae, we find that this great concurrence would take place after an interval of about 1,780 million years. We may notice in passing that this is almost exactly equal to the ages of the oldest radioactive rocks found on earth, and so is probably comparable with the age of the universe itself. This has led to the suggestion that the present universe may be merely the debris resulting from the explosion of a sort of big single super-molecule. But without relying on so realistic an interpretation, we must notice that the mere facts of nebular recession, or even of apparent recession, provide us with a natural unit of 1,780 million years which is of cosmical significance; it has a meaning wherever in the universe we happen to live.

Physics also can provide a natural unit of time—the time light takes to travel across an electron, which is about $1 \cdot 3 \times 10^{-23}$ sec. The ratio of these two natural units of time is $4 \cdot 2 \times 10^{39}$, which is very near to the ratio of the two natural units of length, this being $5 \cdot 0 \times 10^{39}$.

Actually it is not merely by accident that these two numbers are so nearly equal; the theory of the expanding universe shows that the former must be $\frac{1}{2}\sqrt{3}$ times the latter. But, apart from all precise knowledge, anyone who knew nothing of the mathematical theory of the expanding universe might have conjectured that the approximate equality was not accidental. It is strange that the universe should provide two such unequal units of length, and strange that it should provide two such unequal units of time, but it would be strange beyond belief that the units should stand in such similar ratios through mere accident. If we see two seven-foot men coming out of the same house, we shall probably conjecture that they are brothers. On the same principle, we may conjecture that the two large ratios owe their largeness and approximate equality to some common cause, something inherent in the scheme of nature, in which case it would be natural that they should be of the same order of magnitude.

In the case of the units of length and time, this something can be found in the conception of the expanding universe. But we have also found a similar ratio for the two natural units of force, of which the value is $2 \cdot 3 \times 10^{39}$. As far back as 1920 Hermann Weyl of Göttingen suggested that this ratio and that of the units of length might owe their largeness and approximate equality to some common cause. In 1922 Eddington pointed out that this same cause, if it existed, would also make N great, so that it would not be surprising that each ratio should be approximately equal to $\sqrt{N}$.

If we agree that the approximate equality of these large numbers can be no accident, but must result from something in the fixed order of nature, then it ought to be possible to discover this something, show that it accounts for the approximate equality of the numbers, and also calculate the exact values of the numbers. Eddington claims to have done precisely this, and has deduced values for the fundamental physical constants of nature which agree remarkably well with those determined in the laboratory. Unfortunately the details of the problem are far too intricate for discussion here. But, details apart, Eddington introduces us to a universe which is as it is—or rather, of which our picture is as it is—from sheer mathematical necessity; it is as it is because it could not be otherwise without breaking the laws of mathematics and logic. And as evidence of the truth of his theories, he claims to have calculated the true values of the constants of nature from them. On the other hand it should be remarked that the theory leads to a world consisting solely of electrons and protons—the only two kinds of particles which were known when Eddington formulated the theory. It not only failed to predict the existence of the other fundamental particles which have since been discovered, but (so far as I understand it) also suggests that their existence is impossible. This is particularly true of the positive electron, or positron.

Others have ventured even farther along this road. We have seen how the period of 1,780 million years which we extracted from the apparent motions of the nebulae provides a cosmic unit of time, which is comparable with the age of the universe. An alternative procedure would be to take the age of the universe itself as the unit of time. At present this is of the order of 10^{39} physical units of time, but when the universe is 10 times as old as now, it will be about 10^{40} physical units. On the view we are now considering, the other large numbers which are now of the order of 10^{39} will by then also be of the order of 10^{40}. These numbers are all obtained by combining various of the so-called 'constants' of nature, whence it follows that these 'constants' cannot be truly constants, but must slowly change their values. As $\sqrt{N}$ is one of these numbers, it would seem to follow, on the view I am now describing, that N itself is not a constant, but must increase with the time—as though creation were still in progress. This conclusion would violate the general principle of the conservation of mass and would be in even more flagrant opposition with Eddington's hypothesis that N is a mathematical constant, for, whatever else changes, the mathematical constants at least must remain eternally the same. A reconciliation can be arranged by making the unit of time continually change its value, so that the age of the universe in terms of this changing unit remains always the same; we must make a world in which 'it is always six o'clock'. And this is very easily done if we can assume that the constants of nature are themselves capable of change.

Professor Milne has been led, by quite another road, to postulate changes of precisely this kind. Far be it from me to discuss his general work here, on his own native heath, but perhaps I may be permitted to point out its relation to what I have just been saying. Starting from a general hypothetical principle, entirely different from those I have described, he finds it necessary to postulate two distinct kinds of time, one physical and one astronomical. These are measured in different units, the ratio of which changes as the age of the universe increases. At present the ratio of the units is unity because we elect to make it so as a matter of convenience. But it is in process of change through a continual variation of the so-called 'constants of nature', such as the constant of gravitation and the charge on the electron. Actually there are more quantities at our disposal than there are equations to be satisfied, so that there are an infinite number of ways of achieving the necessary result. This only amounts to saying that there are many ways in which the observed phenomena can be pictured consistently.

Now it is, I think, generally agreed that Professor Milne's scheme is simply an alternative description to that of the expanding universe, which again is alternative to the shrinking atom, and again, as Milne has recently shown, is substantially similar to that of the ageing light-quanta. All these descriptions predict the same set of phenomena. If they did not, we could appeal to the phenomena to decide between them, and might even be able to reduce the permissible descriptions to a single one. If so, this one would

368

become a working hypothesis, but would still have no claims to represent the truth; all we could say would be that nothing proved it untrue. But so far we know of no way of discriminating by observation.

Before the coming of relativity theory, the final goal which science set before itself was not to find true descriptions of the phenomena, but true explanations. The various descriptions I have mentioned can of course be treated as tentative explanations, and, if absolute space and time still existed, they would all be different. When lengths can be measured against absolute space, an expanding universe cannot be the same thing as a shrinking atom; when we can set our clocks by Newton's 'equably flowing stream of time', a varying constant of gravitation cannot be the same thing as one that is fixed. But with absolute space and time banished from the scene, it is meaningless to say that one explanation is more true than another; absolute truth makes its exit in company with absolute time and space.

In this way, it seems to me that astronomy has reached the same state as atomic physics. To the physicist, a shower of electrons is sometimes a stream of particles, and sometimes a train of waves; we are no longer concerned with which is the truer concept, but only with which is the more convenient for the problem in hand; we know, moreover, that neither expresses the whole truth. In the same way, to the astronomer, space is sometimes expanding; sometimes it stands still while the objects in it contract; sometimes it is flat and sometimes curved, sometimes finite and sometimes infinite. None of these pictures is true in any absolute sense, because space is a creation of our own minds, and we can make it what we like, but every one provides a convenient background against which to depict the phenomena. So long as we speak and think in terms of Newtonian space and time, astronomical truth is a meaningless concept; we can aspire only to consistency—to an internal consistency between the different parts of our description, and to an external consistency between the phenomena and our description of them. But if we can adjust our language and concepts to the demands of the theory of relativity—replacing points, distances, and times by events, intervals, and sequences—we may expect that our now diverse pictures will all merge into one. Then it will become permissible to think of a 'true' description, although any explanation, except in a purely mathematical form, may be for ever beyond our grasp.